Ernest Favenc

The Secret of the Australian Desert

Ernest Favenc

The Secret of the Australian Desert

ISBN/EAN: 9783743313408

Manufactured in Europe, USA, Canada, Australia, Japa

Cover: Foto ©berggeist007 / pixelio.de

Manufactured and distributed by brebook publishing software
(www.brebook.com)

Ernest Favenc

The Secret of the Australian Desert

THE SECRET

OF THE

AUSTRALIAN DESERT

BY

ERNEST FAVENC

Author of "The History of Australian Exploration", "Tales of the Austral
Tropics", &c. &c.

*WITH FOUR ILLUSTRATIONS BY PERCY F. S. SPENCE
AND A MAP*

LONDON

BLACKIE & SON, Limited, 50 OLD BAILEY, E.C.
GLASGOW AND DUBLIN

PREFACE.

Although the interior of the continent of Australia is singularly deficient in the more picturesque elements of romance, it was, for nearly two-thirds of a century, a most attractive lure to men of adventurous character.

Oxley, Sturt, Mitchell, Kennedy, and Stuart have left deathless names on the roll of Australian explorers, but the unknown fate of Ludwig Leichhardt has always centred most of the romance of story about his memory.

In April, 1848, he left Macpherson's Station, Cogoon River, situated in the southern portion of what is now the colony of Queensland, with the intention of endeavouring to reach Perth, on the west coast, the capital of Western Australia, by traversing, if possible, the then unknown heart of the continent.

From that day to this, no clue to the disappearance of the whole party has ever been discovered. Several expeditions have been fruitlessly despatched in search of the missing men; and many false reports as to the finding of relics of the party have been brought in at various times. Even the rapid advance of settlement, and the comparatively full knowledge now possessed

of the interior, have thrown no light on the subject.
This, the great mystery of Australian exploration, I
have taken for the groundwork of my story.

The view I have adopted of the probable course
pursued by Leichhardt and his party, is the one that
commends itself to the majority of experienced bushmen. Turned back by the dry country west of the
Diamantina River, the explorer probably followed that
river up, and crossed the main watershed on to the
head of some river running north into the Gulf of
Carpentaria; in fact, the same track afterwards followed
by the ill-fated Burke and Wills. Leichhardt could
then easily reach the route he pursued on his first
expedition to Port Essington, the only successful one
he made, and on which his reputation is based. This
course would then lead him around the foot of the
Gulf to the Roper River, where he would leave his
old route, follow the Roper, or a tributary, to its head,
and strike south-west, into the scantily-watered waste
of the interior. This view is borne out by the fact
that trees, marked with what appears to be a letter
L, have been found on or near this supposititious
line of travel; and A. C. Gregory, the leader of one
of the search expeditions, discovered the framework of
a small hut, seemingly built by white men, on a creek
he called the Elsie, a tributary of the Roper River.

Another unexplained riddle I have introduced
points to the possible early occupation of Australia

by an ancient and partly civilized race. In 1838, Lieutenant, now Sir George Grey, when on an expedition in north-west Australia, discovered some remarkable paintings in a cave on the Glenelg River, evidently not the work of the present inhabitants. He describes the principal one as—"The figure of a man, ten feet six inches in height, clothed from the chin downwards in a red garment which reached to the wrists and ankles". The head of this figure was encircled by a halo or turban, and on it strange characters were inscribed, like a written name. Grey says—"I was certainly surprised at the moment I first saw this gigantic head and upper part of a body bending over and staring grimly down on me". Although the dress and accessories so plainly prove that these paintings were not the work of the Australian aborigines, the locality, strange to say, has not been again investigated. I have taken the liberty of transplanting these cave paintings from the north-west coast to the interior, and also of changing the names of some of the members of Leichhardt's party. The descriptions of the physical features of the country are faithful records from personal experience.

ERNEST FAVENC.

SYDNEY, N.S.W., *September*, 1894.

CONTENTS.

Chap.		Page
I.	Sand and Scrub,	11
II.	A Strange Road,	19
III.	A Mysterious Procession,	26
IV.	The Devil's Tracks,	34
V.	A Lifeless Swamp,	41
VI.	The Burning Mountain at Last,	46
VII.	Cannibals,	55
VIII.	The Fight in the Cave,	68
IX.	An Exciting Discovery,	80
X.	The Missing Expedition,	87
XI.	Stuart's Journal,	93
XII.	Charlie's Adventure,	102
XIII.	The Trip South,	112
XIV.	In the Spinifex Desert,	120
XV.	The Fate of Columbus,	128
XVI.	The Slaughter Chamber,	134
XVII.	A Hopeless Situation,	141

Chap.		Page
XVIII.	The Ancient Australians,	149
XIX.	Charlie Falls Sick,	156
XX.	A Further Discovery,	163
XXI.	The Final Departure,	171
XXII.	The Gold Reef Discovered,	180
XXIII.	A Solitary Camp,	187
XXIV.	More Dry Creeks,	196
XXV.	The Last of the Cannibals,	204
XXVI.	A False Alarm,	212
XXVII.	Home Again,	220

ILLUSTRATIONS.

	Page
THEY FIND THE DEVIL'S TRACK ON THE ROCK-PLAIN, .	35
THE DEATH OF DR. LEICHHARDT IN THE DESERT, . .	95
MORTON AND HIS PARTY EXAMINE THE SLAUGHTER-CHAMBER, ...	151
THE LAST OF THE BLOODTHIRSTY WARLATTAS, *Frontis.*	206
Route Map, ...	15

THE SECRET OF
THE AUSTRALIAN DESERT.

CHAPTER I.

The Start for the Burning Mountain—Sand and Scrub.

IT is the beginning of November—November in the Southern hemisphere, not the raw, foggy month of the North—November in Central Australia, where the sun rises hot and red in a breathless morn, and sinks at night in a heated haze, hovering around the level horizon.

It has been a day to doze in the shade if possible, and dream of icebergs. The short twilight is rapidly fading into the darkness of a moonless night. Scarcely darkness, however, for the brilliant constellations of the south and the radiant evening star in the west lend their rays to light up the scene. Under the verandah of a rough hut—mud walls with galvanized iron roof,—three men are sitting indolently smoking the evening pipe that usually follows the last meal of the day.   It is far up in the north of South Australia, in fact almost on the boundary line that divides that colony from its dependency, known as the Northern Territory.   The hut is the principal building on a cattle-station, where, as on most other outside stations, the improvements are of a very primitive kind.   The three occupants of the verandah are—the owner of the station; a young relation staying with him to gain that much-talked-of commodity, "colonial experience"; and a friend, a squatter from a neighbouring run.

"Well," says Morton, the owner, a sun-tanned, wiry little fellow, addressing his neighbour, "what do you say, Brown, to having a look for the burning mountain?"

"Umph!" grunts Brown, who differs considerably in size, owning as he does some six feet two inches of humanity; "isn't this weather hot enough for you without looking for burning mountains?"

"We've nothing much to do for two or three months, and I've made up my mind to see if there's any truth in this yarn the niggers have."

"I never could make head or tail of it," said Brown.

"Nor I," returned Morton; "but although

everybody puts it down as a burning mountain, I am not of that opinion. I have questioned them very patiently, and can only find out that there is a big fire always burning in the same place, but when I ask about a mountain, they say no. None of them have ever been there; they have only heard of it from others, and they seem almost frightened to speak of it."

"They use much the same word for rocks, stones, and mountains."

"Yes; and I think it is rocks that they mean."

"What has your boy, Billy Button, to say about it?"

"Billy comes from a tribe nearly a hundred miles from here. He has heard the yarn, but has never seen any blacks who have been there."

"Let's see. It is supposed to lie rather north of west from here. How far have you been in that direction, Morton?"

"Some fifty miles. It's all scrub and sand. The niggers, however, get across in some seasons of the year, and I think this is the time; there have been plenty of thunder-storms that way lately."

"Well, I'll make one; a little scorching more or less does not matter much up here. You ought to have kept some of the camels back the last time the team was up here."

"Didn't think of it. But I fancy horses will be handier, we have a thunder-storm nearly every day."

"And shall have until we start," replied Brown, "then you see they will knock off at once. How many of us will there be?"

"The pair of us, and — what do you say, Charlie? Are you anxious to distinguish yourself?"

"I certainly hope you won't leave me behind," returned his young cousin, in an injured tone.

"All right. Billy Button will make four, and that will be enough. To-morrow we'll have all the horses in and get ready for a start the next day."

"How long shall we be away?" asked Charlie, who bore upon his shoulders the onerous duties of storekeeper.

"Can't say. What do you think, Brown? Six weeks? Two months?"

"We surely ought to find something in that time, if it's only the remains of Leichhardt."

"Make up three months' rations for four, Charlie; I hate to run short. Lucky we killed the other day, the beef will be just right for carrying."

On an outside cattle-station, where so much camping-out has to be constantly done, the preparations for such a trip do not take long, and the morning of the second day found everything in readiness. Brown had sent over to his place for his own horses, and they started with fourteen in all. Two apiece for riding, four packed

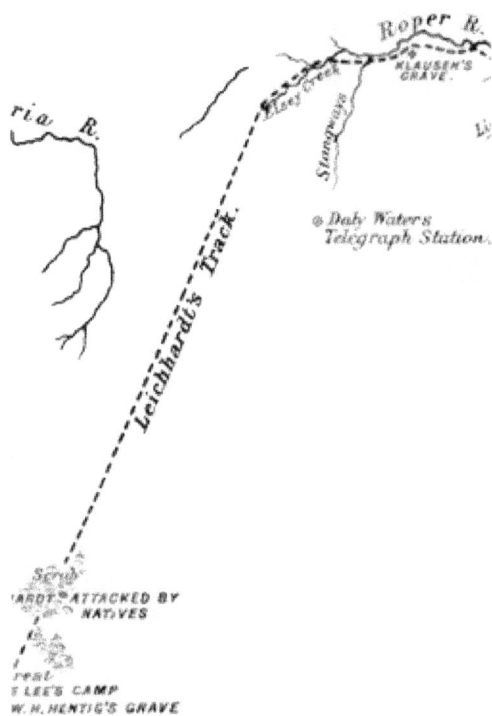

ri.a. R.

Roper R.

KLAUSEN'S GRAVE.

Lees Creek

Stangways

⊕ *Daly Waters Telegraph Station.*

L

Leichhardt's Track.

Scrub

ARDT ATTACKED BY NATIVES

rest

S LEE'S CAMP

W. H. HENTIG'S GRAVE

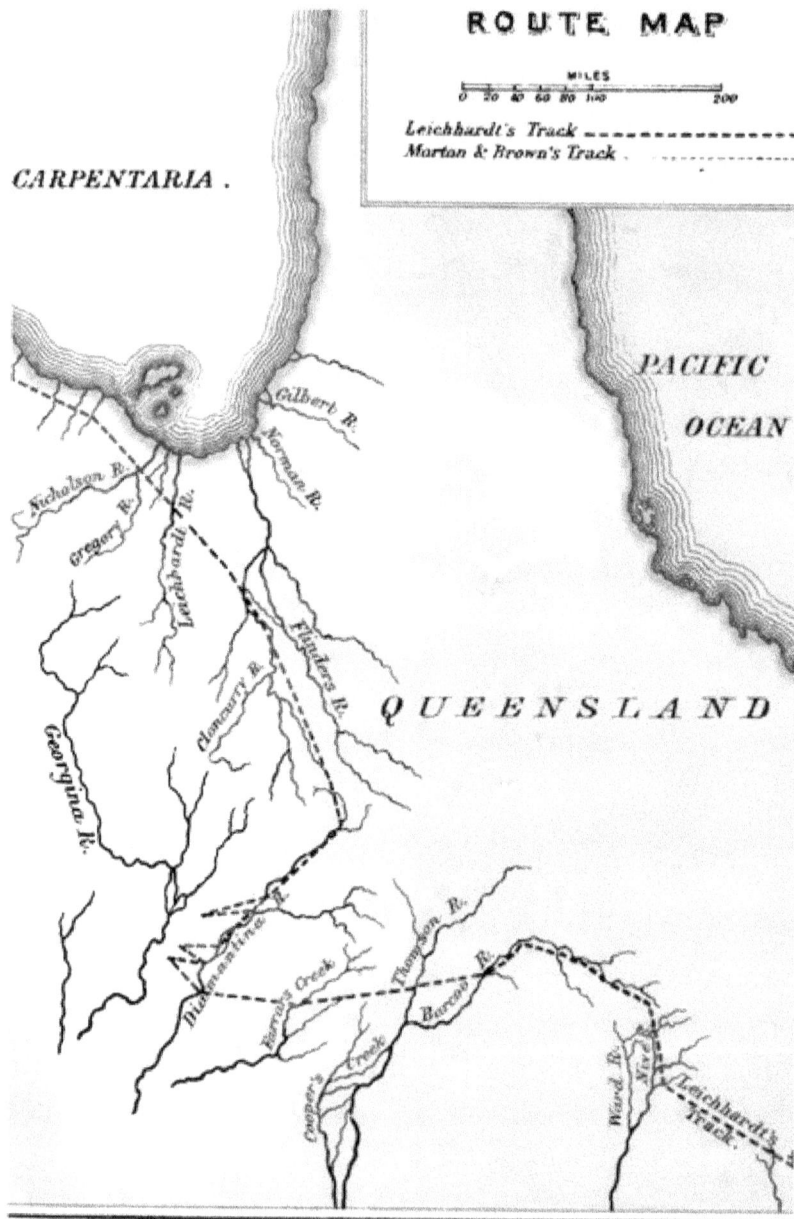

ROUTE MAP

MILES

| 0 | 20 | 40 | 60 | 80 | 100 | | 200 |

Leichhardt's Track _ _ _ _ _ _ _ _
Morton & Brown's Track ....

CARPENTARIA.

PACIFIC

OCEAN

Gilbert R.

Norman R.

Nicholson R.

Gregory R.

Leichhardt R.

Cloncurry R.

Flinders R.

Georgina R.

QUEENSLAND

Diamantina R.

Thomson R.

Barcoo Creek

Barcoo R.

Cooper's Creek

Ward Riv.

Leichhardt's Track.

SNOW & T. DUBLIN.

with rations, and two with canvas water-bags and the necessary blankets, tent, &c. At the last moment the blacks about the station tried to dissuade Billy from going by telling him horrible tales of the fate surely awaiting him at the dreaded burning mountain, but Billy stoutly refused to be frightened, and scorned to remain, although given the option by Morton.

The first thirty miles of the journey was over familiar country, and they camped that night at a small water-hole lately filled by a thunder-storm. Beyond them now stretched a waste of sand ridges and mulga scrub, into which Morton had once penetrated for some twenty miles. With full water-bags, and a determination not to be beaten back without a struggle, our adventurers commenced the second day's journey with light hearts.

During the whole of the day the sombre scrub and heavy sand continued, without break or change in their depressing monotony. Scarcely the note of a bird or insect broke the silence, as they toiled on without much heart for conversation. Towards evening a piece of good fortune befell them. On a small flat between two sand ridges they crossed a patch of short green grass, the result of a recent thunder-storm. No water could be found, the hot summer sun having evaporated all that had been caught in the shallow clay pans. The green grass was, however, a boon to the horses, who did not feel the want of water so much on the soft young feed.

Next morning they were saddled and packed up, ready to start by sunrise. About ten o'clock they ascended a sand ridge somewhat higher than those they had formerly crossed, and from its crest they were able to look around on the sea of scrub that surrounded them. Not far off, in the direction in which they were going, Morton drew the attention of his companions to a thin column of smoke.

"Burning mountain already?" queried Brown.

"Niggers travelling and hunting," replied Morton. "The scrub looks thinner there. They won't be far from camp at this time in the morning; but I expect the water is only a soak-hole, of no use to us."

In less than an hour they were riding over patches of still-burning grass, thinly scattered through a forest of bloodwood-trees; but neither the sharp eyes of Morton or Billy could detect a sign of the hunters. After searching for some time the boy found the tracks of a blackfellow, two gins,[1] and some pickaninnies[2] coming from the westward, and these they followed back for about a mile to a freshly-abandoned camp. It was situated on a fairly open piece of country, partly covered with coarse drift sand. Not far from the camp was a ragged old shell of a gum-

---

[1] Women.   [2] Children.

tree, covered with tomahawk marks. Billy, who had at once gone to this tree, gave a low whistle, and the others came up. He pointed to a small hole near the butt, and dismounting put his arm down and then peered into it.

"Water long way down," he said. "Gone bung, mine think it." By which they understood that the supply had dried up. After some searching about, a long sapling was procured and thrust down. The hole was about ten feet deep, and the end of the sapling brought up some wet mud.

"How did the blacks get down for the water, Billy?" asked Brown.

"Pickaninny go down," replied the boy, pointing to a tiny foothold in the side of the hole.

"Well, boys," said Morton, who had been poking the sapling down vigorously and examining the point, "I don't see much to be got out of this. Evidently there's been one little family living on this hole, and now they've been dried out. It would take us two hours to open up this hole, and then we should probably get nothing for our pains."

"Water gone bung," repeated Billy.

"What do you say to following this flat? It's going partly in our direction, and may lead to something."

No one having anything better to suggest they resumed their journey once more, until a mid-day halt was made.

(M 64) B

"Well, respected leader," remarked Brown, after the meal was finished and pipes were lit, "I'm afraid our horses will look mighty dicky to-morrow morning unless we get them a drink to-night."

Morton glanced lazily at them, where they stood grouped under whatever scanty shade they could obtain.

"They are beginning to look tucked up," he replied, "but we'll pull up something before dark."

"I sincerely hope so," said Brown as he stood up. "Go ahead once more, Captain Cook."

About four o'clock the open flat which they had followed grew narrower, until at last the scrub closed in entirely and they found themselves confronted by a thicker growth than any they had yet met with. The mulga having given place to a species of mallee.

Morton, who was leading, stopped.

"We must push through," he said. "It may be only a belt, and if we start to follow it round we shall be all night in it."

"Right," replied Brown. "I'll take a turn ahead if you like. I prefer being first in a scrub."

Morton laughed and dropped behind, and for about an hour very slow progress was made, the scrub getting worse and worse. The sun was sinking low, and the cheerful prospect of a night in the scrub was before them, when, to the relief of all, Brown suddenly called out:

"Hurrah! we're out of it!"

CHAPTER II.

A Native Cemetery—Billy's Explanation—Stopped once more by Dense Scrub—Discovery of a Strange Road.

AS the party emerged, one after another, from the scrub, their eyes were delighted by a prospect of open-downs country before them, dotted here and there with clumps of gidea-trees. But, better than all, there was plainly to be seen, scarcely a short mile away, a line of gum-trees, creek timber, whilst the presence of water was plainly attested by flights of white corellas hovering about.

It was not long before the whole party were comfortably encamped beside a good-sized water-hole, and the horses luxuriating on succulent Mitchell and blue grass.

Brown, with his pipe as usual under full blast, was enjoying the scene, when Billy, who had been wandering around the camp, came up and re-marked:

"No sleep here."

"What's the matter?" asked Brown.

Billy pointed to a patch of scrub a short distance off, and beckoned to him to follow.

Brown noticed that the tops of the trees looked particularly thick and dense, but it was not until he was quite close that he saw the reason. Nearly every tree of any size bore a rude scaffolding, and on the top of every scaffold lay either a bleached skeleton or a dried mummy-like corpse. The ground, too, was covered with bones and skulls that had fallen through. Brown called the others, and they gazed with awe at this strange sepulchre.

"I've often seen the bodies put in trees, but never in such numbers as this. Why, there must be hundreds here!" said Morton.

"I never saw more than two together at the outside," returned Brown. "Strange," he went on, after a closer inspection; "all the bodies who have any dried skin remaining on their foreheads have a red smudge there!"

"No sleep here; by and by that fellow get up, walk about," insisted Billy.

This remark helped to dispel the gloom caused by the sight of so many dead bodies, and Billy had to undergo a good deal of chaff. It was evident, however, that his fright was genuine, although, like most natives, the reason of it could not be drawn from him.

No ghostly visitants came near the camp that night, and all slept the sleep of tired men.

Charlie, waking up before daylight and finding Billy in the sound stupor common to the aborigines at that hour, conceived a wicked idea. Brown dabbled a little in sketching, and Charlie, after hunting up the colour-box in one of the pack-bags, proceeded to paint Billy's forehead red, after the manner of the mummies in the tree-tops.

"Hallo, Billy!" said Morton, when they were all about and the quart-pots for breakfast merrily boiling. "What's up with your head?"

Billy grinned, not understanding what was meant.

"Look here," said Brown, taking a hand-glass out of the pack and holding it in front of his face. Billy looked, and turned as white as it was possible for a blackfellow to do.

"Him bin come up!" he yelled, starting up and pointing to the scrub where the bodies were. Then looked apprehensively around, as though he expected to see some belated corpse still walking about.

"Tell him you did it, Charlie," said Morton. "I'm afraid you've funked him, and if so he'll bolt. Never play tricks on a blackfellow."

Charlie at once complied, and after Billy had been induced to wash the paint off and had inspected the colour-box, he was somewhat comforted; but he evidently still thought that the subject was not a fit one to joke about.

Struck by Billy's evident panic, Morton again

attempted to extract the reason from him, and
after some trouble learned that he had heard of
the men with a red smear on the forehead, who
were supposed to be in some way connected with
the burning mountain. That, during the daytime, they pretended to be dead, but at night
got up and walked about.

"This looks as though we were on the right
track," said Morton to Brown.

"Hum! Nice sort of company you are intro-
ducing us to. However—Death or glory! Let's
saddle up and make a start."

In a short time the friendly water-hole and
the ghastly scrub beside it were left behind,
but the patch of open country unfortunately
proved to be of very limited extent in the direction they were going, and in a short time they
found themselves again entangled in the dense
scrub, which was now becoming such a formi-
dable obstacle to their progress. Towards the
middle of the day, the sanguine Morton began
to despair of pushing on, even at the slow rate
at which they were going, and to meditate a
return to their last night's camp and a fresh
start in a new direction. At noon they were
compelled to halt; the desert hedge-wood had
now made its appearance, and the barrier presented by it was almost impenetrable.

They stopped for a hasty meal, and when it
was finished, Morton said to Brown:

"What do you say, old man? Will you go north for a bit, and I will go south, and we'll see if there is anything like a gap in this confounded scrub?"

"My dear old boy I am entirely at your disposal. But allow me to suggest that we shall get along infinitely better on foot."

"I think so too. Charlie, you and Billy stop here with the horses until we come back."

It was a good two hours before the cracking of branches and muttered bad language, coming from the south, announced to Charlie and Billy the return of Morton.

"How did you get on?" was the query.

"Get on!" returned Morton savagely; "I did not get on at all. I don't believe I got half a mile from here. It's the worst old-man scrub I was ever in in my life; I've barked my hands nicely. If old Brown did not get on any better than I did, we shall have to go and chop him out with an axe."

Almost as he spoke, Billy held up his hand and said:

"Mitter Brown come up."

In a few minutes his tall form emerged from the thicket.

"I beg to report, sir," he said to Morton with mock solemnity, "that the main road to somewhere is about three-quarters of a mile to the northward."

" What on earth do you mean, old man?"

" Just what I say. After fighting my way through some of the most awful scrub I ever met with, I came to a fine clear road—gas-lamps, milestones, and probably bridges and public-houses."

" Well, we'd better go there at once. I wonder you came back without patronizing one of the pubs."

" I did not exactly see all that I have stated, but I have no doubt whatever of their existence," returned Brown. " Joking apart, there really is a cleared track out there, but we'll have to cut a road to get the horses there."

" This bangs everything into a dust-heap. But it's getting late and we had better shape. Charlie, you and Billy go ahead with the tomahawks, and we will dodge the horses along after you."

It took time, labour, and patience to make the distance indicated by Brown; but about an hour before sundown, to the astonishment of three of them, they stood upon what was evidently a cleared track, about the width of an ordinary bridle-track. Morton examined the stumps, and pointed out that the work had been done by stone tomahawks. Billy looked for tracks, but none had been made since the rain from the last thunder-storm had fallen.

" It's running westward. I suppose it's all

right to follow it, but this sort of thing beats my experience. What say you, Brown?" asked Morton.

"Forward, gentlemen, while the light lasts," was the reply of that individual.

Their progress was now easy, for the track had been most carefully cleared, and the horses, all old stagers, marched along in single file without any trouble. Darkness, however, fell, and the scrub was still on either hand of them unchanged.

"Morton," said Brown, breaking the silence, "I've got an idea."

"Stick to it hard, old man; it's the first I ever knew you to possess."

"Don't try to be too funny. Well, I shouldn't be in the least surprised to meet a first-class funeral coming along at any moment."

"You're worse than Billy."

"Billy was partly right. Those old mummies, skeletons, &c., we saw back there, have all been carted along this road from—wherever we're going to. That is the reason it is so carefully cleared."

"Jove! your right. And we might have come along this road all the way if we had kept our eyes open, instead of tearing ourselves to pieces in the scrub, travelling parallel with it."

"That view of the question did not occur to me, but it's a perfectly feasible one."

"Rather a surprise for the mourners if we blunder on to them in the dark to-night."

"Just what we want to avoid. There's something ahead no white man has yet heard of, and if we can sneak along without our presence being suspected, so much the better."

"What do you propose? We can't budge a step off the track just now, and if unluckily there happens to be a funeral ceremony on to-night, there's bound to be a collision."

"We must go on until we come to a piece of open country, and then pull off and wait for daylight."

"All serene. But our tracks will tell tales."

"We can't help that, unfortunately."

The conversation had been carried on without halting, and the march now continued in silence, until a low whistle from Morton gave the signal to pull up.

CHAPTER III.

A Midnight Halt—A Mysterious Procession—Sudden Dispersion and Flight—Open Country once more and another Mystery Ahead.

AS well as could be made out in the gloom cast by the scrub, they had reached a small break in it, and Morton wheeling off, the others followed, and the party dismounted, as the leader judged, some two hundred yards from the track. Morton gave his orders in low tones, for the atmosphere of awe and mystery affected everybody. There was no grass, so the horses were simply relieved of their packs and tied to trees; then the men stretched themselves on their blankets without making a fire, and, save for the occasional stamp and snort of a horse, the scrub was as silent as before the white men roused the echoes.

Not for long.

It seemed to Brown that he had scarcely closed his eyes when the camp was aroused by a distant melancholy cry. No one spoke; all were too intently engaged in listening. The cry sounded again, louder, nearer, and in a chorus of many voices.

"What bad luck," whispered Morton to his friend. "One day sooner or later and we would have been right."

Nearer and nearer came the plaintive wailing, and the gleam of firesticks was visible. It was a most uncomfortable sensation that our adventurers experienced, lying prone and motionless in the gloomy scrub listening to this weird procession passing through the desert land. They were well armed, and confident against any number of aborigines, but the sights they had encountered were so much out of the ordinary bush routine as to make even such old hands as Brown and Morton feel slightly nervous. Charlie was naturally much excited, while Billy was "larding the lean earth" with the perspiration of abject, superstitious fear.

The party of natives were now opposite to them, and not very far away, and by the number of firesticks they judged that there must be a good many in the company. Every now and then the wild wail or chant kept breaking out, and the shuffling noise of their bare feet was distinctly audible during the silent intervals.

They had almost passed the hidden watchers, when the procession was interrupted by a sudden and discordant shout from the leaders. A babble of voices followed, the firesticks gathered together for a moment, and were then dashed on the ground and extinguished. Next came the noise of feet flying back along the track; these died rapidly away in the distance, and the scrub was as silent as before.

"Saw our tracks!" said Brown with a disgusted sigh, breaking the spell that held them all quiet.

"How could they see our tracks in the dark?" asked Charlie.

"They could both feel and smell them," returned Morton. "The ground is caked hard from the last thunder-storm, and our horses walking one after the other have cut it up soft. Of course, with their bare feet they could tell the difference at once. The scent, too, would be as plain as possible at this time in the morning, even to one of us. What's the time, Brown?"

Brown struck a match.

"Three. It will be breaking day soon after five. Let's wait till then."

"Why?" demanded Morton. "We might as well get along while it's cool. There's the remains of a moon just rising."

"Why? Because you think with me that it was a funeral party. Now, I should like to know what they did with the body; they never carried it away with them at that pace."

"Never thought of that," returned Morton. "Yes, we might pick up some information by waiting until daylight and seeing what they threw away. Make a fire, and we'll have breakfast."

The time soon passed in discussing the strange scene just witnessed and the probable result of their trip. Morton reminded Brown of the freemasons M'Dowall Stuart asserted he met with amongst the aborigines in the interior, and Charlie, who had not heard the former conversation, was enlightened as to the probable meaning of what had just passed.

As soon as daylight was strong enough the investigation commenced. Right on the track where it had been hastily dropped lay the dead body of a man. A tall old man, fastened on to

a rude litter of saplings. The forehead was
smeared with red pigment, and on the dusky
breast was a triangle inscribed in white.

Brown gave a low whistle.

"That's a thing I never saw blacks draw
before," he said to Morton.

"Nor I.   He's a fine-looking old boy.   What
a long white beard he has got for a nigger!"

The corpse was fastened to the litter with
strips of curragong bark; and they were turning
away after noticing these details, when Brown
suggested that they had better move it off the
track.

"You know," he explained, "we might come
bustling back here in a bigger hurry than those
fellows were, and tumble over the old gentleman
in the dark."

The litter and its burden were shifted a few
paces in the scrub, and, full of expectation, the
party resumed their interrupted journey.

The break where they had halted was the
beginning of the outskirts of the scrub; the
country soon became more open, and as it did so
the track they were following grew less marked.
It was still, however, quite plain enough for any
bushman to follow easily.   At noon, to the great
relief of the horses, they came to a small pool of
rain-water, and some fairly good grass.   Here
they turned out for a long spell.

"Question is," said Brown, when the usual

discussion commenced, " Where did those nigs camp? No sign of them here. By the way, Billy, did you notice any gins' tracks amongst them?"

" No," returned the boy. "Altogether black-fellow."

" Must be more water ahead; and I hope so, for this won't last another week, and we want something permanent to fall back on. Now, I'm going aloft on the look-out," said Morton.

Charlie watched him curiously as he slung the field-glass over his shoulder, and taking a tomahawk proceeded to an exceptionally tall blood-wood-tree near the camp. At the foot he took off his boots, and cutting niches in the trunk, as a blackfellow does when climbing, he was soon up amongst the topmost branches. Ensconcing himself firmly, he took a comprehensive sweep around with the glasses, and then directed his attention to the westward.

" Below there!" he shouted, after a lengthened scrutiny.

" Hi, hi, sir!" returned Charlie.

" Brown! Will your long legs bring you up here safely?"

" Well, I'll try." And in a short time Brown was up alongside his friend, and a very earnest discussion followed, extremely tantalizing to Charlie down below. After taking a compass-bearing to some distant object they descended;

and Charlie, who was already barefooted, immediately attempted the ascent, slipping ignominiously down after getting up two or three steps, to the intense delight of Billy. With the black boy's assistance, however, and much sarcastic advice from his cousin and Brown, he managed to reach the first branches, and thence easily gained the perch Morton had occupied on the top.

What did he see when he got there?

To the westward the forest soon came to an abrupt stop, and beyond stretched a great gray plain, bounded by something that Charlie could not make out, and which had evidently puzzled Brown and Morton. It was not water, although it looked something like it; it was a broad sheet of pale blue, glistening in places under the sun's rays, and beyond, above a quivering haze, was a dark object like a distant ridge.

"What name, Billy?" said Charlie to the black boy, who had climbed up after him. "Water?"

"*Bal*," said Billy decidedly. "Water sit down here, close up," he added, pointing to the edge of the forest.

"What name, then?" repeated Charlie.

"Mine think it mud, where water bin go bung," was the blackfellow's opinion, and with this they both descended.

"Well, Charlie, what do you make of it?" asked Morton,

"Billy thinks it's mud where the water has dried up," returned Charlie, as he had no opinion of his own to offer.

"And Billy's right, I believe. It must be the bed of a dry salt lake; but we'll get along to the edge of the timber and camp."

On the margin of the plain they came to some fine lagoons, with good grass for the horses, but nothing could be seen of the mysterious object ahead, excepting from the top of a tree.

On the banks of the lagoons they found abundant traces of the natives, and it was evidently a main camping-place on their way to and from their burial-place. Many of the trees were marked with triangles, a sign which considerably puzzled the elder travellers. The open country, the ample supply of water, and the relief from the gloomy surroundings of the scrub had restored the cheerful tone of the party, and imparted a sense of security to them.

But neither Brown nor Morton were men to neglect due precautions, now that their presence was known to the probably hostile inhabitants. So a watch was kept all night by the three whites in turn, Billy escaping the vigil, as blacks are not to be trusted to keep awake.

CHAPTER IV.

The Limestone Plain—The Devil's Tracks—A Strange Mark.

THE morning found them early on the move, the night having passed without any alarm, false or real. They still followed the faint track leading straight toward the dark ridge they had seen beyond the blue expanse. This supposed dry lake had been visible from the camp before sunrise, but as the sun rose it disappeared, nor did they again sight it until nearly eight o'clock. At ten they were close to it, and all doubt as to its character was set at rest. They pulled up, not at the edge of a dry salt lake, but of an unbroken sheet of limestone rock. Nothing was visible ahead but this stony sea of bluish-gray, over which a heated haze was undulating. The dark line beyond, resembling a ridge, had vanished, and the wind that blew in their faces across the surface of this strange plain, was as hot as though it came from the open door of a furnace.

Morton turned and rode along the edge of the rock to where the pad came in, for they had left the track for the last hundred yards. He whistled, and the others joined him. The track

THEY FIND THE DEVIL'S TRACK ON THE ROCK-PLAIN.

still continued right on across the rock, but its course was now indicated by other means. On the surface of the limestone had been scratched and chipped with infinite care, an imitation of human footsteps, or rather more than human footsteps, for the gigantic tracks were more than twice the size of a man's, and a stride to correspond was indicated. Side by side, about six feet apart, these two awful footsteps disappeared into the quivering mirage.

"I've seen that mark before on the granite mounds in Western Australia," said Brown. "You notice that there are six toes to it. It's supposed to be the footprint of the devil."

"By Jove, what tedious work it must have been cutting those marks!" returned Morton. "They're not lazy beggars ahead of us whatever else they may be. But what shall we do now?"

"Go back to the lagoons. It's a rattling good camp, and we have heaps of time before us. We'll hold a council of war this afternoon and decide upon some course of action."

"Right," answered Morton. "We shall have to go slowly and cannily or we shall be getting into a tight place."

They returned to their former camp, and, as evening drew on, entered into a discussion as to their immediate movements.

"Brown, you're the longest, speak first," said Morton.

"Those beggars are located beyond that lime-stone rock. Is not that so?"

"Yes."

"They may be the most mild and peaceful people going, and they may be the most trucu-lent ruffians. I incline to the latter opinion."

"So do I, but I cannot say why exactly. They took to their heels quick enough the other night."

"Oh, any niggers will do that on a sudden start. However, it's safest to act as though they were our enemies."

"Decidedly."

"To-morrow we'll go right and left along the edge of the rock for a few miles on each side of the track, and see if there's any other track they use. If there's only the one, why, we know where to expect them from."

To this Morton agreed, but suggested that two should follow the track across the rock.

"No, old man, you're too eager," said Brown. "We're too small a party to afford to split up. When we go across that rock we must all go, and take pot luck."

"You're right," agreed Morton. "To-morrow you and I will go along the edge of the rock. Charlie, you and Billy will stop and mind camp and examine all the trees about for marks, in fact have a good fossick round."

"When we cross the rock we shall have to go

on foot, we can't take the horses across," said Brown.

"Certainly not, and I doubt if we can cross on foot in the daytime. We should be baked to death with the rock underneath and the sun overhead. We should get no shade to rest under the whole way across."

"The horses will be safe enough here while we are away. If the niggers use only the one track, why, we are bound to meet them."

Another quiet night was passed, although a watch was kept. In the cool morning Brown and Morton started across the plain, leaving Charlie to scour about the camp. Billy, arrayed in a light and airy costume consisting of a saddle-strap and a tomahawk, had evidently laid himself out for a day's pleasurable sport.

"This plain seems fairly well grassed," said Morton as they rode across. "Wonder how far it extends?"

"We'll find out before we get back. But country is not of much value out here just now, no matter how good it is."

"No, worse luck; you and I know that to our cost."

When they reached the rock they separated, Brown going north and Morton south. Following the edge along, without going into all the dips and bends, Morton went on until he reckoned he had covered some six miles. The limestone

rock pursued much the same course to the southward, but the forest and the continuation of the
chain of lagoons at its edge bore in towards the
rock, and it was evident that the two would
meet in time.

Morton rode over to the edge of the timber,
and found that the water-course there was still
well supplied with occasional pools of water.
He could see no tracks of blacks there, nor were
there any marks on the rock: all was lifeless
and lonely, save for the tireless kites. As he
rode back, however, he caught sight of a bird
high up in the air steadily flying to the west.
He recognized it as an eagle-hawk, and was
astonished to see others following, all flying in
the same direction. Then the discordant note of
a crow came to him, and a flock of the black
creatures flew past, conversing in the peculiar
guttural croak common to crows when on the
wing. They, too, were going across the rock to
the westward.

"Hang me, if there isn't a rendezvous over
there somewhere of all the carrion birds in the
district," said Morton.

He rode on and found Brown at the meeting-
place, he having got back sooner. His experience
had been somewhat similar for the first few
miles; then the country changed, a low stunted
forest obtruded from the east, and the ground
became hard, stony, and barren, save for patches

of spinifex.[1] The limestone rock, too, became
more uneven and broken, and it was evident
that he had approached the verge of the formation they were then traversing; probably, he
thought, the change would result in a large expanse of desert, spinifex country.

"We could get round that way," he remarked,
"without having to cross this rock."

"Better stick to the track; then we know we
are going straight to wherever these triangle men
came from," replied Morton.

"Did you see any niggers' tracks?"

"Not a sign of any. I don't think I saw or
heard a living thing of any kind since leaving."

Morton told him about the flight of hawks
and crows he had noticed, and as they rode back
to camp they decided to make an excursion to
the south before tackling the great rock and the
mystery beyond. It was as well to know all
about the country before making their final start;
and, moreover, if the natives came back and saw
their tracks going away south it might throw
them off their guard.

Charlie and Billy had found nothing about
the camp beyond a peculiar mark cut on a tree,
which somewhat differed from the others they
had seen. They had caught some fish in the
largest of the lagoons, and Billy had a fine big
carpet snake roasting in the ashes; for no matter

---

[1] A wiry, prickly grass, useless as fodder.

how well fed a blackfellow is, he always likes to revert to his aboriginal delicacies occasionally.

Charlie took them to inspect the new mark, which was on a large flooded box-tree. It had been chopped out with a stone tomahawk in the rugged bark, and must have taken much time and labour. Both men looked at it from all points of view without arriving at any conclusion; then, just as they were turning away, Morton exclaimed:

"If I saw that tatooed on a sailor's arm I should say that it was meant for an anchor."

The two others instantly recognized the resemblance, and they all came to the conclusion that it was a rude attempt to depict that emblem.

"Mystery thickens," said Brown. "Are we going to reach the much-talked-of inland sea and find a race of sailor-men in possession?"

"Devilish queer," replied Morton; "it seems to me the sort of mark an illiterate white man who had been a sailor would make on a tree. It's chopped out very neatly, much as a sailor would do anything of the sort."

"I suppose we shall find out all about it before we get to the end of this trip," returned Brown.

"Yes, and a good deal more than we now dream of, I anticipate."

"Did you have a good look to the south and north when you were up that tree?" asked Brown.

"No, I didn't. My attention was at once taken up by the strange-looking rock ahead of us."

"So was mine. I think we might go up another at sundown; we might see something."

When the sun nearly touched the horizon they ascended the tallest tree in the neighbourhood, but nothing was discernible southward. To the north, however, a low range was visible a long distance away.

A quiet undisturbed night succeeded, and an early start was made the next morning.

CHAPTER V.

Hot Springs—A Lifeless Swamp—More Marks of the Natives.

THE first six miles being over the country traversed by Morton was naturally uninteresting. Then the plain grew narrower and narrower. The chain of lagoons where they had camped developed into a large water-course, and the flat limestone rock began to alter its character and soon merged into a basaltic ridge coming from the westward. At mid-day the plain was a thing of the past, and they were now travelling along a broad water-course, with open forest on one side and a rude line of basalt boulders, piled like a wall, on the other.

At a fair-sized lagoon, thick with water-lilies, they turned out for their meal.

"Funny," remarked Brown, "how these inland rivers disappear. This water-course looks big enough now, but I bet it runs out to nothing before night."

"Yes; the wet seasons, I suppose, are very rare, and when one comes the flood-water is absorbed by soakage and evaporation before it can cut a continuous channel. You know that no rivers enter the sea to the south of us."

"I know; it's all a wall of cliffs around the head of the Great Bight. Was there not some yarn once about fresh water being obtained there some distance at sea?"

"I've heard something about it; it was put down to the discharge of a subterranean river, but I don't think the fact was ever proved."

"Well, if we find a river of that sort we'll make a canoe and send Charlie and Billy down it to explore. What do you say, Charlie?"

"There might be some Jinkarras living down there," replied Charlie.

"Ever see any Jinkarras, Billy?" asked Morton.

"No. Plenty bin hear 'em," replied Billy.

"I wonder how this yarn of an underground race, the Jinkarras, originated."

"I can't make out. The noise they hear at

night that they say is made by the Jinkarras is made by a bird—a kind of quail."

"Well, we must be off; pack up boys," said Brown.

About four o'clock a dense mass of foliage was visible ahead, which, as they drew nearer, proved to be huge paper-bark-trees, with long trailing branches, like gigantic weeping willows. The ground around these ancient giants was soft and spongy, and the bed of the creek was soon lost. The ground being too soft to allow of the horses progressing any further, a camp was made and they were hobbled out.

Leaving Billy to light a fire and mind camp, the three whites went on foot through this great white forest. The ground grew swampier as they proceeded, until at last, when within sight of a belt of tall reeds, they could proceed no farther. Moreover, the water was getting uncomfortably warm.

"Hot soda springs," said Brown; "this accounts for the growth of these trees. There's an easy one to climb," pointing to a bending one. "Let's go aloft and look ahead."

The tree was easy of ascent, and the three were soon high up amongst the branches. Beyond the reeds lay a lakelet of clear water, but, save for the deep fringe of rushes, not a plant of any sort was visible. No ducks or other aquatic birds could be seen.

"I guess that water's too warm for anything to live in it," said Morton.

It was a strange scene; the sun was sinking low, and anywhere else the place at that time would have been busy with feathered life, but here all was lifeless. The lakelet, surrounded by its border of tall reeds, in which there was apparently no break, lay there calm and unruffled.

"Let's get back to camp," said Brown. "Looks as though we'd got into a dead corner of the world."

Next morning it was determined to follow the swamp round to the westward to ascertain its extent. In a mile or two they came to where the basalt wall apparently ran out in the swamp, disappearing in a few scattered boulders. Just beyond this they came to a well-beaten track, which came round the swamp from the direction in which they were going and turned off amongst the basalt. Following this track along, in about a mile they came on two skeletons lying beside it. Some dry bits of skin still adhered here and there on the fleshless bones.

"A nice part of Australia this," remarked Brown, as they halted and gazed at the poor remains. "If we're not falling foul of a cemetery or a funeral, we run against skeletons lying promiscuously about. Wonder what brought them here."

"Not want of water at any rate," replied Morton. "Been a fight, perhaps. They don't seem to belong to those triangle gentry, at any rate."

"They must have been lying here months," answered Brown; "they're past our help anyway. May as well get on."

Gradually the swamps rounded off to the southward, and it was evident that there was no continuation that way. The creek they had followed from the lagoons had disappeared, as Brown had predicted. To the south was nothing but a stony, desert forest of stunted trees, the ground covered with spinifex. Strange to say, however, the track that had partly circled the hot swamp had branched off and headed due south. There had been some discussion as to whether they should follow it or not, but, as it was evident that this track and the one marked by the devil's footsteps were trending to the same centre, it was decided to postpone it until they had solved that mystery first. The swamp was of such a circumference that it was nearly sundown when they got back to the site of their last night's camp, having crossed no outflow from the swamp anywhere.

In the morning it was thought better to go back to the old camp at the lagoons and follow the devil's footsteps, than try to follow the track amongst the boulders coming from the south.

They arrived there early, and immediately made

preparations for burying their spare rations, ammunition, &c. The next few hours were busy ones. Saddles, rations, spare ammunition, &c., were all carefully buried, and the whereabouts masked by a fire being lit on the top to hide the disturbance of the earth. They started as soon as this work was finished, each man carrying rifle, revolver, ammunition, three days' rations, quart-pot, and water-bag—a fair load for men always accustomed to riding only. Most devoutly they all prayed that they would be off the rock soon after daylight the next morning.

CHAPTER VI.

The Night March across the Great Rock—Meeting with the Natives—The Secret of the Burning Mountain.

WITH all the despatch they made they did not reach the edge of the rock before twelve; but it mattered little, as the surface was only then getting cool, and would have been unbearable any earlier.

Billy was sent first with bare feet, he being trusted to follow the track by feeling when they strayed off it, as he would then cross the rough surface made by the sculptured footprints, the remainder of the limestone being almost as smooth as marble.

It was a weird and weary tramp across this rock by the light of the stars, with vague darkness all around them. None of them felt inclined to speak, and an intense silence reigned everywhere. A sickly moon rose just before daylight, and its faint beams cast the long shadows of the travellers across the gleaming surface of the limestone. The thought in the minds of the three white men was the same—what would daylight show them? Billy plodded along mechanically; most of the time he was half asleep.

Daylight came at last, and the black line that they had seen from the tree-top gradually came into view, apparently not far ahead; and each felt grateful that he had not to encounter the force of the sun on the face of the naked rock.

When it was broad daylight the dark line resolved itself under the glasses into a row of basaltic boulders, with some bushes growing in their clefts and a bottle-tree here and there on their summits.

"We shall be there before it is hot," said Morton thankfully, as he closed the glasses; "let's push on."

They did so, and before the sun had attained much power found themselves amongst the boulders. The track led straight into a gap, and on one side a huge block of stone, supported by two others, made a rude cave, under which the weary men gladly took shelter after their toilsome tramp. Evidently it was a halting-place for the blacks, for the remains of fires were about, and a supply of firewood, which came in very handy for the tired men to cook their breakfast with.

A satisfactory meal and a smoke being finished, the situation was reviewed. Behind them lay the bare expanse of rock just crossed, and before them the unknown. Now, too, they would have to keep a keen look-out for lurking foes, because in amongst these boulders every step was fraught with danger, especially as the blacks knew of their approach; and it was evident that they were trespassing on tabooed ground. The future movements of the party were now, as might be supposed, a matter of serious consideration, and Brown and Morton were in earnest discussion when a loud report like a clap of thunder suddenly startled the little company to their feet. A low rumbling followed that seemed to shake the very rocks. Hurrying outside, nothing was seen that could possibly have caused the strange noise; the sky was cloudless, the air still and sultry.

Suddenly Billy pointed to the westward. "Fire jump up," he said.

A puff of white smoke, or vapour, was rising, seemingly only a short distance from them. Silently they watched it ascend and disperse.

"Blacks here have artillery apparently," said Brown. "Salute in honour of our arrival."

Nothing more following, they returned to the cave, leaving Charlie at the entrance on the look-out.

"If these fellows know nothing about the effect of firearms," said Morton, "we may be able to establish a funk; they may have heard of them only from the other tribes."

"I don't think they have much communication with the other tribes by the look of it, but, if they live amongst these rocks, what on earth do they exist on?—for there's no game here."

"Well, all we can do is to keep a sharp look-out and our powder dry—What's up?"

"Here's the corpse!" cried Charlie, falling back from the entrance in amazement.

Billy gave an awful yell; the others started to their feet as a tall native coolly walked into the cave, and squatted down on the ground. It certainly was enough to give them all a fright, for the visitor, in outward appearance, greatly resembled the dead man left in the scrub. A second glance, however, showed points of difference, which proved him to be a denizen of the earth; he was marked with the white triangle on the breast, and the red smear on the forehead, but was naked and unarmed, whilst his manner showed no trace of fear. Recovering themselves somewhat, Morton lugged Billy forward to see if he could converse with the new-comer. This proceeding, however, did not suit their visitor, for he addressed a furious tirade at Mr. Billy Button in some unknown tongue, winding up by violently spitting at him.

Billy slunk back scared, and the native, rising, took Brown by the arm and led him to the entrance; pointing alternately forward and backwards he made signs for them to turn back, and not go on. Brown returned answer by signs that they must go on. The blackfellow shook his head vigorously, and then held up his hand motioning them to listen. Again the loud report was heard, and a puff of vapour ascended as before.

To the apparent surprise of the native, the whites showed no alarm, and Brown taking his carbine stepped back, and fired it into the air. The black gave a decided start, and trembled a little, but stood his ground; then his mind seemed to change, and, making a sign to Brown to stay, he strode off and disappeared behind the surrounding rocks.

" Is he coming back, do you think?" said Brown.

"I think so," replied Morton. " He's a fine fellow, with plenty of pluck."

"Then we'll give him twenty minutes' grace —but here he comes, and all his sisters and his cousins and his aunts."

Sure enough their former visitor appeared, accompanied by some half-dozen others, similarly painted and all unarmed. He spoke a few words to them, and pointed towards Brown, upon whom they gazed curiously.

"Now then, Brown," said Morton, "you're the star; evidently they want an exhibition."

Brown, who had reloaded his carbine, fired it in the air again. The fresh arrivals showed more alarm than the first man had, some of them squatted hastily down and all started with fear.

"It appears to me, Brown, that they consider you 'brother belonging' to this noise ahead," remarked Morton.

"It looks like it, and we must keep up the pleasing fiction, for these fellows 'have us on toast' in amongst these rocks. I wonder how many more there are round about."

"Let's see if we can go on now," said Morton.

On Brown indicating their wish to proceed, the most ready acquiescence was displayed, and at a few words from the native who had first arrived the others showed by signs their intention to carry the strangers' packs.

Before starting, however, names were interchanged, that being generally found the easiest steps towards intimacy. Brown, Morton, and Charlie (or Sarley) were soon picked up, and the chief, as he appeared to be, introduced himself as Columberi, which, of course, was at once turned into Columbus by Brown, and the oldest blackfellow amongst the others was named Yarlow.

With Columbus appropriately in the lead, the

march commenced, the tracks winding in and
out among the rocks in a very intricate fashion.
For nearly five miles they kept on, although in a
straight line they would not probably have
traversed more than two, and at last arrived at
an open space surrounded by bottle-trees, and
from the number of humpies,[1] built of mud and
grass, apparently the head-quarters of the tribe.
Here they halted. About twenty more blacks
were sitting about, who at first made a show of
taking to their heels, but a call from Columbus
brought them back.

Selecting a shady tree, Brown indicated that
their swags should be brought up, and this being
done he remarked:

"What do you say to a feed, and then getting
Christopher here to show us the ropes?"

"Just as well," returned Brown; "we must take
everything as a matter of course, and show no
surprise."

Billy made a fire, and the quarts were put on
to boil, a proceeding which interested the spectators greatly. Brown by signs then invited
Columbus to sit down, and presented him with
a piece of damper thickly coated with sugar, at
the same time eating a piece himself to inspire
confidence.

The native started to eat in a slow and doubt-

[1] A native hut built of bark or sticks plastered with mud:
called also a "gunyah" or a "whirlie".

ful manner, but after a bite or two finished it off with great gusto and indicated a wish for more.

The quarts now bubbled up, and the blacks with one accord emitted a united "ha!"[1] and pointed to the westward; evidently the boiling water bore some resemblance to something in that direction.

Columbus now described a mark in the dust like a half-circle, and pointed in the direction they had come. "He means the horse-tracks they saw," said Morton, after a pause. He nodded vigorously to the old man, who continued his pantomime by lying on his back as though dead. Morton nodded again and patted the ground, pointing backwards to indicate that the corpse was still there.

Columbus then called the other blacks aside, and after a long talk half a dozen of them drew off and disappeared amongst the surrounding boulders.

"We must follow up this burning mountain business," said Brown, as soon as he had eaten his dinner; "now old Columbus has disposed of his private affairs perhaps he will take us there."

"Call him back and let's make inquiries; see if he'll eat beef."

The chieftain approaching, Brown offered him a piece of salt beef. He examined it curiously,

---

[1] Wild Australian blacks know nothing of boiling water. They make water hot by putting hot stones in it.

and then without any demur ate it in the most
appreciative manner. He then pointed to Charlie,
and made signs as though cutting with a knife,
which for a time were quite unintelligible.

"Blessed if he doesn't mean to ask if you're
good to eat," said Brown at last.

He shook his head, and the native appeared
both surprised and disappointed. On their indicating their wish to proceed in the direction
of the strange reports, he rose and led the way.
The whites only took their carbines, as they felt
assured that as yet their coming was too novel
for the blacks to interfere with their belongings.

They had but a short distance to go. Round-
ing a rugged wall of basalt they saw stretched
before them a singular and striking scene. At
their feet was a large circular shallow depression,
about a mile in circumference, filled with pools
of clear water divided one from another by
narrow ridges of rock. In the centre of this
depression was a hill of small elevation with a
flat top; not a vestige of green weed was to be seen
about the water, nothing but bare rock. Without
stopping, their guide led the way along one of the
narrow strips of basalt intersecting the water.

"Keep your feet," said Morton, as they followed
him, "for it strikes me that water's scalding hot."

Warned by this, the whites carefully continued
their course to the central hill. Columbus
mounted it, and then pointed down. They were

on the edge of a crater. At no great distance below was a mass of seething boiling mud. The crater had lip-like fractures in various parts, and to one of these their guide now directed their attention, at the same time motioning to them to stand back from the edge. The water in the pool at the back of the lip was curiously ruffled; presently it assumed the appearance of boiling, and, rising suddenly, poured over the edge of the crater into the molten mud beneath. A deafening report followed, and the rocks on which the party crouched trembled again. Then came a rush of steam, and all was still once more. By a great effort the strangers had preserved their coolness, and looked on the display unmoved; then, in response to Brown, they discharged their carbines simultaneously, an act which nearly made Columbus topple into the crater.

CHAPTER VII.

Columbus takes a Fancy to Charlie—The Secret of the Limestone Cliff—A Feast of Cannibals—A White Man with the Blacks—Initiation of Recruits—Charlie makes a Proposition.

BY the language of signs they were given to understand that the rocks through which they had found their way extended in every direction. Another low elevation a short distance away resembling a limestone cliff was noticed, but about this their guide, who had now recovered his composure, could, or apparently would, not afford them any information. After a more lengthened examination of the strange surroundings they returned to their camp in the open space, which they found deserted by the natives. Columbus, however, showed no signs of leaving them, and the whites, with due regard to strategic purposes, pitched their tent and made themselves as comfortable as circumstances allowed.

"The thing that puzzles me," said Brown, after all arrangements had been completed, "is —What do these natives live on? Columbus, whom we have feasted on strange dainties, shows no desire of leaving us; but the others are all away, evidently in search of grub. There are no gins visible; perhaps they are away hunting, but I doubt it, for within a hundred of miles of here there isn't a feed for a bandicoot."

"I don't understand it either," returned Morton, "but we'll stop and see it out, anyway. Charlie, our friend Columbus has taken quite a fancy to you; he can't keep his eyes off you.'

Charlie looked very uncomfortable at the chaff, and muttered something about a "nigger's cheek"; but it was quite evident that the native had transferred all his admiration from Brown to Charlie.

Whilst still talking and discussing the situation a sound like a distant uproar of voices became apparent, and Columbus commenced to evince signs of uneasiness. The sound came from the direction of the limestone cliffs, and grew louder and more distinct as they listened. All the party naturally rose to their feet, although the native made energetic signs to them to keep quiet. After a short time the shouting became stationary, and it was evidently not intended as an attack upon them, or such loud warning would not have been given.

"Shall we go and see what's up?" said Morton.

"We'll fix the direction, anyway," returned Brown, and they proceeded to clamber up one of the high boulders by which they were surrounded, although Columbus evidently protested against the proceeding.

From the top of the boulder they could make out the summit of the limestone cliffs, and ascertained that the uproar certainly came from there, and, moreover, that the shrill cries of gins mingled with the many voices. It was well on towards sundown, and after a short conference Brown and Morton determined to defer further explorations until the next day, so they returned to their camp.

Columbus, who seemed much relieved by the

proceeding, now made signs for Charlie to accompany him in the direction they had been just looking. At the same time he made it plainly apparent that Charlie was to come alone.

"I'll go, Frank," he said to Morton. "Let me go and have all the honour and glory."

Morton and Brown both replied in the negative, and Brown intimated to Columbus that to-morrow Charlie should go, but now it was nearly night and he wanted a sleep. This seemed to satisfy the blackfellow, who evidently wanted to get away himself, and presently, as soon as he thought the attention of the party was not directed towards him, he disappeared as mysteriously as he had arrived.

"I have not got to the bottom of this little affair yet," said Morton; "but I think we shall to-night. What do you say to paying a visit to these cliffs as soon as it is pitch dark—I have the bearing?"

"The very thing I was going to suggest. Charlie, it strikes me that our new friend wants to make long pig of you."

"What's that?" asked Charlie.

"Well, a favourite dish amongst some natives who have an acquired taste for human flesh."

"Do you think he's a cannibal?" said the boy, rather aghast.

"I should be sorry to slander a stranger, but it certainly looks something like it."

As soon as it was quite dark the party set out on their way to the cliffs, which they judged to be about a mile distant; it was a difficult matter shaping a course by the stars amongst the gloom cast by the surrounding boulders, but an occasional murmur of sound helped them on, and after scrambling and twisting about they found themselves near the low cliffs. Here Billy was told to strip and reconnoitre, and his black figure was lost amongst the rocks almost before he seemed to have made a step. He was absent nearly half an hour; then a subdued whistle announced his return, and in a low voice he communicated to Morton the result of his investigations.

About four or five hundred yards from where they were waiting there was a cave in the cliff, and the blacks it appears were in there. Billy had gone close to the entrance, but could see only a light in the distance, for, according to him, "hole bin go long way".

Under Billy's guidance they soon reached the cave entrance, and found it to be a kind of tunnel evidently leading to a large cave, for a red glare of firelight came round an angle, and the sound of many chattering voices was audible.

"Shall we go on?" said Morton in a whisper.

"No, wait a minute," replied Brown; "it strikes me there's another entrance to this place; they must have a lot of fire going, but yet the place is not full of smoke. I can smell the fire, but that's all. I think there must be an opening in the top; let's send Billy up to see."

The face of the cliff was easily climbed, being mostly detached rocks that had fallen down, and very soon Billy came back and reported that "fire come up alonga top".

One after the other the adventurers ascended, and found themselves on a rocky plateau full of fissures and holes, through some of which a bright light was streaming. Approaching this portion carefully on their hands and knees, they soon found a fissure through which they could gaze with safety on one of the strangest scenes ever witnessed in Australia.

The cavern below them was seemingly of some size, and was well lighted by a number of fires, the smoke from which somewhat annoyed the unseen spectators. A far larger number of blacks were assembled than had been visible before, and many of them were armed and painted, being also marked with the red smear and white triangle. One large group was composed of some twenty or thirty young men and women; they were huddled together, apparently much frightened, and had no marks whatever upon their bodies.

Columbus was soon recognized squatting at one of the fires with some of the other old men, and, like all but the group of boys and girls, busily engaged in eating. Morton felt his arm clutched suddenly and tightly, and Brown hoarsely whispered in his ear:

"It is meat they are eating; but what meat?"

Morton was struck with horror as he listened, and the truth flashed across his mind. It was a feast of cannibals they were overlooking. The armed natives had just returned from a foray, and the trembling group in the corner were prisoners destined to death.

An awful feeling of horror came over the whole party as they realized their situation and possible fate. In a wilderness of savage rocks, surrounded by an expanse of desert, almost in the hands of some fifty or sixty fierce cannibals —no wonder the first impulse of each was to slip quietly back the way he had come under cover of the night, and leave the natives to their former obscurity.

Their natural audacity, however, soon returned. At present they were masters of the situation; with their breech-loaders they could shoot down a score of the natives, helpless in the cavern below, if so inclined. But, with all their horror of the scene, affairs did not seem to justify armed intervention just then.

Contenting themselves with being spectators only, they watched the doings in the cave, at times having to stifle a cough brought on by a puff of smoke from the burning wood fires.

For a time the repast below went on with the usual accompaniments of a blacks' camp, but as it came to an end it was evident that some extraordinary occurrence was going to take place. Gradually the old men mustered together around Columbus, and the other blacks proceeded to combine all the fires into one large one near the wall of the cavern. The added blaze gave to view a huge figure on the rock; it was the semblance of a human form, but the head, instead of being represented round, was grotesquely shaped liked a triangle. At the foot of the painting was a rock, and while the rank and file of the natives grouped themselves in a circle around the fire, Columbus and some others retired into the darkness out of sight of the watchers. The chant of a *corroboree* now commenced, and the blacks slowly circled round the fire for a short time, suddenly ceasing and breaking into a half ring, with the open part towards the grim figure painted on the wall. Then Columbus and the others appeared, supporting between them a striking and venerable figure— an old, old man, with snow-white hair and beard, bent so double that, as he hobbled along supporting himself on two short sticks, he appeared like some strange animal walking on four legs. This decrepit being was carefully helped and guided to the stone beneath the figure, and seated thereon; then the others squatted on the ground, the blacks in the half ring remaining quietly standing.

The old man seated on the block was now full in view of the whites above, and the brilliant rays of the fire fell directly on him. Brown and Morton turned to each other with the same smothered exclamations on their lips:

"By Jove, it's a white man!"

Almost as dark as the savages around, painted like them with a hideous red smear on the forehead and a white triangle on the breast, the experienced whites yet felt sure that before them they saw one of their own race. Apparently the venerable being was either blind or nearly so, and he kept turning his face restlessly from side to side. From the half circle of blacks then arose a shout or chant that sounded like the repetition of "Mur! Fee! Mur! Fee!"

"Hullo! we're amongst countrymen," whispered Brown; "that sounds awfully like Murphy."

A terrible noise now commenced like a hundred mad gongs let loose. Four blacks came forward, beating furiously with clubs on what appeared to be sheets of metal. At the sound the old man on the rock smiled and leant forward, and, stretching forth his trembling hands, appeared to say something.

At this Columbus arose, and, followed by the gong-beaters, went over to the throng of trembling captives. After a short inspection he

selected a young gin and pulled her along by
the hand towards the old man, followed still by
the gong-beaters. The poor wretch seemed stu-
pefied with fear, and when in front of the stone
she sank down, trembling visibly. Columbus
drew back, and the gong-beaters, dancing madly
round, made a still more deafening din. Suddenly
one of them, instead of striking his gong, dealt
the unfortunate creature a terrible blow on the
head, the other gong-beaters followed his example, and in an instant the wretched gin lay dead
on the ground.

The effect of this scene on the whites above
was maddening. Charlie had his gun to his
shoulder, but Morton stopped him in time.
The gin was killed before interference was
possible.

"Come away," said Brown; "let's have a con-
fab. I'm sick of watching those brutes."

They scrambled away a short distance, and
after a pause Brown spoke.

"We've got our work cut out, there's no doubt
about that. We must find out all about that
white man if possible, and we must release those
poor devils and give these cannibals a lesson."

"In justice to our friend Columbus," said
Morton, "let me remark that 'these cannibals'
are only following up what they have been taught;
they have no horror of the thing like we have.
At the same time, the man who lifts his hand

—or nulla nulla—against a woman is unworthy the name of a British sailor, &c. &c."

"Are you convinced that that is a white man?" said Charlie.

"Yes," replied Brown; "but who he is is another question. He appears to be blind, deaf, and imbecile. I suppose we must fall back upon Leichhardt."

"He's been a big man when younger, and erect," said Morton; "far bigger than Leichhardt was. However, we'll suppose it to be one of his party—he looks old enough."

Brown gave vent to a low whistle.

"By jingo, supposing that was 'Murphy' they were shouting. I believe there was a man of that name in the lost party."

"We shall find out, I hope, soon. Meantime, what next?"

"I know," said Charlie; "let's go back to camp. You promised Columbus I should go with him to morrow. Well, I'll go and find out all about it."

Morton put his hand on Charlie's shoulder.

"It's a real plucky bid, my boy, after what we've just seen; but do you think I could let you go? Why, you'd be cooked and eaten in no time."

"Hold on!" said Brown; "I'm full of ideas just now; let me think this one out; there's something in what Charlie says."

" Now, oracle, as soon as you're ready," returned Morton.

"Well, I may be right or wrong, but my notion is that Columbus does not want to eat Charlie. Why, they've got enough rations for a month. I think that they keep this old man as a sort of Fetich, and that Columbus and a few of the knowing old fellows see that he must soon die. Now they want Charlie to take his place."

Brown paused triumphantly.

"I verily believe you've hit it," said Morton. " You ought always to live here, considering the amount of intellect you are developing."

At this moment a renewed din once more pealed up from the cave, and the party crawled back to find out the cause. The gong-beaters, Columbus, and his privy councillors were parading the captives, and the spectators shuddered as they looked down upon hideous remains of the late feast scattered about the sandy floor of the cavern. This time a fine-looking young man was selected and marched up to the venerable figure on the stone. The gong-beaters fell back, and Columbus and his companions proceeded to smear the youth's forehead with red pigment, and marked the cabalistic white triangle on his breast. He was then led away to a dark corner out of sight of the watchers.

Brown muttered a deep oath.

"That's what has been puzzling me," he growled to Morton, "how they kept their numbers up; of course they recruit from the best-looking prisoners."

"See! they are going to select another," whispered his friend. "Bet you two figs of tobacco they choose that tall fellow with his hair tied in a knot."

"Done! I'll back the nuggety fellow alongside him."

Brown lost, the tall fellow being marched out to receive the marks of the cannibalistic brotherhood.

Columbus and the others now assisted the old man to hobble away, and the blacks squatted down by the fire and relit fresh ones about the ground.

"Get back to camp," said Morton; "the circus is over for to-night."

Scrambling down the cliff, and using every precaution, the party soon regained their camp, which they found as deserted as they left it.

CHAPTER VIII.

Charlie as Decoy—Death of the old White Man—The
Fight in the Cave—The Catastrophe.

TIRED out with their exertions and the continued night work the party slept soundly, and awoke at dawn to find the camp as calm and silent as if no such tragedies as they had witnessed were ever enacted in the neighbourhood.

"Terribly sultry, is it not?" said Morton; "I suppose it is because these rocks retain the heat so."

"It seems in the air. Look what a haze there is. I don't think I ever felt it so hot at this time of day. What do you say to a walk to the crater after breakfast?"

Charlie called out just then that the meal was ready, and during its progress the plan of action for the day was discussed and agreed upon.

On arriving at the crater they found it in a great state of activity, most of the pools were in violent commotion, and constantly overflowing into the crater, causing a succession of reports.

Returning to the camp they found that Columbus and two or three of the old men had arrived, all looking as mild and gentle as if they habitually lived upon milk and water.

"Look at the old scoundrel," said Morton;

" his mouth is watering to see us roasting on the coals."

"I think he only wants to get rid of us and to induce us to leave Charlie behind. Now let's try him," returned Brown.

Preparations were then apparently made for departure, Charlie intimating to Columbus that he intended to go with him. The native appeared hugely delighted, and when the time for departure arrived neither he nor the others could restrain their expressions of joy. With their swags on their shoulders Morton, Brown, and Billy strode off along the track by which they had come, ostentatiously waving their hands to Charlie.

No sooner, however, were they hidden from the camp than Morton and Billy slipped aside amongst the rocks, whilst Brown plodded steadily on, making as much noise as possible. For nearly a quarter of a mile he kept his course, and then stepped on one side and stood quietly behind a boulder. After five minutes' waiting the sound of footsteps was heard, and a native came along, evidently following the track to make sure that the white men left the place, for he was unarmed and alone. He was close up before he saw Brown, and then with a frightened cry sprang away; but he was too late. Brown had hold of him, and exerting all his uncommon strength threw him heavily down amongst the rocks,

where he lay stunned and quiet. Brown waited
patiently for some time, but nothing could be
heard; evidently one native only had been sent
to watch them away.

Leaving his swag in a secure hiding-place,
Brown then cautiously directed his course towards
the limestone cliff, using every precaution to
escape being seen. He arrived in sight of the
mouth of the cave after a toilsome journey, and
after cautiously reconnoitring gave a low whistle.
There was no answer, but voices could be heard
approaching, and peering carefully out Brown
saw Columbus, Charlie, and three other old men
emerge from among the rocks and enter the
cave. At the same moment a low whistle sounded
near him, to which he instantly replied, and in
a few minutes Morton and Billy came creeping
silently along and joined him.

"It's splendid," said Morton. "Columbus took
Charlie on one side amongst the rocks, then he
gave a signal and all the blacks came along the
track and squatted down in the open space where
we were camped. Columbus and three old men
then went away with Charlie, whom they carefully kept hidden, and I think those are all we
have to deal with; so come along, for I can't bear
to let that boy stop alone with them long,
although I think he's safe enough."

"We'll just rush the four of them, and then
take our time examining the place and the white

man—is that it?" said Brown; "but how we're to get away afterwards I can't make out."

"We must trust to chance and our rifles; I think we can manage. But come quick."

Noiselessly they stole along the narrow entrance that led into the inner cave, and cautiously peered in to be sure of their ground before making their attack; the prisoners were there and the three old men, but Columbus and Charlie were absent. "Quick!" whispered Brown, and sprang forward on to one, while Morton felled the other with a nulla nulla[1] he had picked up. The third made a bolt for the entrance, uttering a shrill yell as he did so, but Billy, whether through sudden fright or not, fired his carbine at him and the black dropped dead.

Startled by the yell and report, Columbus came rushing from a dark corner of the cave; his eyes were flashing, and all the cannibal in his nature seemed aroused.

"Hit this fellow on the head!" roared Brown, releasing his struggling prisoner and grappling with the new foe.

Morton dealt the native a stunning blow with the waddy,[1] and then turned to assist his comrade. Strong as Brown was, it would have been hard work for him to subdue the infuriated Columbus without assistance. Between them they got him down and bound him with straps.

[1] A native club of hard wood.

"Now for Charlie!" cried Morton, turning in the direction of the dark corner. "Something must have happened to him."

"I'm all right, old man; come with me." And Charlie showed himself at the entrance of another and inner cave.

First stopping to tell Billy to wait and watch the prisoners, and shoot them if they attempted to escape, the two friends followed their young companion, leaving a strange scene behind them —Billy Button on guard at the entrance of the passage, the savages prostrate on the ground, and the captives for the cannibal feast, who had preserved a frightened apathy throughout, still huddled together.

In a smaller cave than the one they had just quitted, lighted like it through fissures from above, the three whites found the old man seated on the sandy floor, gazing with his half sightless eyes at the unaccustomed figures, for thus much could he apparently discern. In a hasty whisper Charlie confided to them that he had been speaking to him, and thought he could make him hear.

"Try again," said Morton eagerly.

Charlie stooped down and shouted in the old man's ear, "Englishman! White man!"

A faint gleam of intelligence seemed to illuminate the poor creature's face, and he pointed eagerly forward with trembling hands. The friends followed the direction of his hands, and saw a heap of objects piled in a dusky corner of the cavern, and Brown strode forward to examine them. The attention of the other two was confined to the ancient white man, who seemed strangely moved. He tried to rise and speak, but could only struggle ineffectually. It was awful to watch his convulsed features, and think what secrets he carried hidden in his breast, secrets that time had forbidden him to reveal. At last with panting effort he half rose up, and with a quavering hoarse voice cried distinctly:

" Yes! Englishman! White man!" and with a choking gasp fell back dead.

Awe-struck and startled the whites looked at the body of the unfortunate man who had dragged out such a long term of existence amongst savages. Not a doubt was in their minds but that they were gazing on one of the survivors of Leichhardt's lost party, whose fate had long been such a mystery. Now the very shock of their coming seemed to have shaken the last sands of life out, and he had died before their eyes with the story of the past untold.

" Look!" said Charlie, stooping gently over the body and indicating the swarthy breast.

There, almost undecipherable by reason of the darkened colour of the skin, was the tatooed mark of a rude anchor.

Suddenly their meditations were interrupted by a series of frantic yells from the outside cave,

and the report of a rifle. Rushing out, the cause
was instantly explained. Billy's attention had
wandered to one of the lady captives, and
Columbus had, unobserved by him, freed himself
from the hastily-tied straps. The first thing
Billy knew about it was a blow from a club,
and the back view of a figure flying up the
entrance-passage, at whom he hastily and vainly
fired, as was pretty evident by the fierce shouts
of Columbus outside, calling his comrades to
him.

"Get you're cartridges ready; we must fight
for all we're worth," cried Brown.

Almost as he spoke there was a rush of flying
feet and a roar of voices at the entrance.

"Fire like blazes!" ordered Morton, setting an
example, which was followed by the others, until
the white smoke nearly filled the cavern. Madly
and fanatically the natives dashed up the narrow
passage; but with four breech-loaders playing on
them, the terrible, unknown lightning and deafening thunder smiting their foremost down, two
and three at a time, the attempt was hopeless;
they fell back, and for a moment or two there
was silence.

"Top! Top! Look out!" suddenly screamed
Billy, and none too soon.

Clambering up the cliff the blacks were on the
plateau forming the roof of the cave, and were
forcing their way down through the many cracks

and fissures. Hastily abandoning their position, the whites had scant time to escape into the open air over the bodies of those they had shot down. Here, to their astonishment, they found themselves unopposed by the cannibals, who had all made for the top of the cliff to gain entrance into the cave.

"What's up, Brown?" cried Morton; "you look like a ghost! Are you hit?"

"No, I don't think so, but I feel queer, and you look sick. For Heaven's sake come over to the rocks, quick!"

An awful feeling of nausea and giddiness suddenly and strangely attacked them all. Reeling to the rocks in front of the cavern they threw themselves down in what shade they could find, utterly regardless of their enemies.

The air was pulsating with fiery heat, the reports from the crater followed each other with scarce any interval, and the earth seemed rocking beneath them. From the mouth of the cavern issued a melancholy wail, the death-chant over the dead white man.

By a great effort Morton rallied himself, for it suddenly flashed across him what was going to happen.

"Come on!" he shouted, staggering to his feet and making to where an overhanging boulder afforded some slight shelter. With difficulty the others followed him. As they crouched down, completely unmanned, they felt the ground tremble violently; then came a terrific report, as if the very rocks were rent asunder, and the air was filled with blinding steam and scalding mud.

Dead silence reigned for nearly ten minutes, then Brown gave a deep sigh and raised his head.

"All aboard!" he cried out. "Anybody hurt?"

One by one they answered, stood up, and looked around.

"Pretty warm while it lasted," said Morton; "that's an experience one does not get every day. Those fellows in the cavern were best off."

"Were they?" cried Brown excitedly. "Great Scott! Look there!"

He pointed to the brow of the cliff, and they all saw what had happened. The mouth of the cavern had disappeared, and the shape of the cliff was changed. The earth-tremor they had just experienced had brought down the roof of the cave, and their late enemies and their wretched captives lay buried beneath countless tons of rock.

The death-wail they had heard had been the death-wail of a whole tribe. The cannibals and their victims were in one common tomb.

"And the secret of that white man lies buried there too," said Morton, after a long pause.

"No, I hope not," replied Brown. "I brought

something away from that heap the old man pointed to;" and from the bosom of his shirt he drew out an old-fashioned leather pocket-book.

No one was anxious to examine the contents just then; they were all in a hurry to get back to camp and quench their thirst, and away from the scene of their late adventure. No apparent change had taken place in the surroundings of their camp, and they made a fire and sat down to rest and eat.

"Poor old Columbus!" said Morton. "I cannot help feeling sorry for the old ruffian. He was a real plucky fellow. Do you remember how coolly he walked in to us the morning we got here?"

"Yes, and after all we had no business—according to their ideas—to interfere with their little rites and ceremonies. They treated us in a friendly fashion."

"After all, however, things turned up trumps for us. We would not have had the ghost of a show in a fight amongst these boulders. No, we must thank that earth-tremor for being alive now."

After their meal was over and the four somewhat rested, Morton proposed a stroll to the crater to see how it had fared, for not a single report had been heard since the one accompanying the eruption of mud.

A wondrous change had taken place, they

found. The crater, or what they had taken for one, had subsided, and over its site now flowed an unbroken sheet of water. The mud on the boulders and the turbid condition of the water were the only signs of the late convulsion of nature.

"And so," said Brown, "the burning mountain, such as it was, is gone for good, and we are the only white men living who have seen it, who will now ever see it."

"That's so," commenced Morton, when he was interrupted by a footstep from behind. They all turned hastily.

Scarred, bleeding, and burnt, a most miserable object, there stood Columbus, the only survivor of his tribe. He looked abjectly and imploringly at the whites—apparently it was to their power he attributed the disaster that had happened,—and came forward with a crushed and broken air, gazing woefully at the space where the crater had been.

Brown beckoned, and the blackfellow came up to them.

Just then Charlie and Billy called out loudly that the water was sinking. It was true: the muddy water was rapidly falling, and a whirl-pool was forming in the middle, as though some cavity in the earth had been opened by the late convulsion. Silently they all watched the water as it swirled round quicker and quicker, and a harsh scream went up from it. In less than half an hour the hole was empty, save for a misty vapour that arose. This cleared away, and the bottom of the hole lay bare—a chaotic jumble of boulders coated with mud, and in the centre a dark rift, as though the crater formation had sunk down bodily.

"Anyone feel inclined to go down there?" said Morton.

"Not just at present," replied Brown; "we'll let it cool off a bit first."

The disappearance of the water seemed to put the final blow on the shattered Columbus. He followed them readily to the camping-ground, where they gave him some food, which he ate ravenously, although it made the whites shudder to see him, when they remembered what his last meal had been.

In spite of what they had gone through, they were all too anxious to get out of the gloomy desert of barren rocks to defer their departure, and at sundown they started back for the lagoons. The ex-chief accompanied them, as they thought they could make him useful in furthering their future discoveries.

They arrived at their camp early the next day tired out, but right glad to get back to more cheerful surroundings. Their horses were feeding quietly about the place, having enjoyed a better time of it than their masters, and everything else was just as they had left it. They endeavoured to extract from Columbus the story of his escape, and after much misunderstanding managed to worry out, that when he found the white man dead he thought that the other white men had killed him, and rushed out after them. As soon as he got outside he was struck down and knew no more, excepting that all the others must have been buried under the fall of rock.

"How about those fellows who were sent back after the corpse?" suddenly said Morton.

Further questioning elicited from Columbus that six men had gone back, and by the signs he used it was evident that they had not yet returned.

"By Jove, I never thought of them!" said Brown. "Lucky they did not come along and spear our horses while we were away."

―――――

CHAPTER IX.

Deciphering the Contents of the Pocket-book—An Exciting Discovery—Another Survivor of Leichhardt's Party, perhaps still living with a Tribe to the Westward—Charlie makes another Proposal.

AS their camp was in every way a good one, and they wanted leisure to decide on their future movements, they determined to remain where they were for a few days.

Brown and Morton set themselves to sort out the contents of the old pocket-book, and Charlie and Billy went fishing and shooting, diversifying their sport with attempts at teaching Columbus to ride.

The pocket-book was found to contain many pages of faded writing, which would evidently take some time to decipher. Some parts were still legible enough, others had suffered mutilation and damage from water and smoke. Fortunately the handwriting appeared to be that of an educated man, so that once they got accustomed to it they would be able to piece it together with a fair amount of ease.

It took them nearly all day to sort the leaves out into the proper sequence of dates, and in doing so they gained a rough idea of the contents. They found that the journal was written by one of three survivors of Dr. Leichhardt's party, named Stuart. He and two others (Kelly and Murphy) had been living for some time with a tribe of friendly blacks to the westward. Kelly had been killed during a fight with the cannibal tribe whose annihilation they had witnessed. The journal recorded up to the death of Kelly and a few weeks beyond, but gave no clue to the subsequent life or fate of the survivors. One of them, Brown and Morton agreed, was the old white man who had died in the cave, but they did not believe that he was the writer of the journal. It was more likely to have been written by Stuart, and the fate of this man greatly excited their curiosity and sympathy. Was he still living with the friendly tribe to the westward?

This question, they felt with sorrow, must be answered in the negative. The presence of his companion, the old white man, evidently a prisoner amongst the cannibals for years, and the strangely preserved unfinished journal, pointed conclusively to another fight, the probable death of Stuart, and the capture of Murphy.

"But," suggested Morton hopefully, "those captives they brought in possibly came from this friendly tribe, which proves that they are still in existence. Why should not Stuart be yet amongst them?"

"I hope so, but cannot think it likely," said Brown. "What sort of a man should you think him to be by the rough idea we have of his journal?"

"A good, self-reliant man."

"Exactly. And I think that if he was still alive he would have trained his tribe up to fight these cannibals, and probably have wiped them out before now and rescued his comrade."

"I must confess that your reasoning sounds conclusive enough, but I won't give up the hope of finding him alive."

"Nor I, although it is hoping against hope."

" We must try and find out from Columbus whether this last batch of victims came from Stuart's tribe; he might know whether he is dead or alive."

That evening Columbus, who had had several spills during his riding-lessons, much to Billy's delight, was interrogated about the tribe to the westward. It came out that there were two tribes which the cannibals harassed, one to the south and one to the west. To the north Columbus intimated that there were no natives. The last raid had been made on the tribe to the westward, who lived by a lake. Further examination elicited the fact that Murphy had been brought from there a long, long time ago; also, that another white man was there who had killed a lot of cannibals and frightened them; but that was also long ago, and now they had been there two or three times and not seen him. They learnt also that he went about everywhere, for he was with the tribe to the south one time when they went there, and had killed some of them there. The southern tribe lived near a mountain.

This was the extent of the information which, after much puzzling on both sides, was gleaned from the cannibal chief.

It rather complicated matters. Was Stuart to the west or south? Which way would they go first? On going into the subject again it appeared

that the way to the south was the easier; to
the west, as was evident by Stuart's journal,
a long stage of dry desert country had to be
crossed.

"At any rate," said Morton, "we have a couple
of days to think it over. We must make a legible
transcription of that journal, and I propose that
we make two copies; I will keep one, Charlie
another, and you, Brown, stick to the original.
This will ensure us somewhat against accident."

"Can I go and explore that hole where the
crater disappeared while you're busy at that?"
said Charlie.

"Go by yourself?" asked Morton.

"No.   I'll take Billy and Columbus."

"And supposing these missing men, the six
sent to take the corpse to the cemetery, turn up
while you are in amongst the rocks?   What
chance would you and Billy have, especially if
Columbus went over to their side?"

"I'd take care Columbus didn't turn traitor,"
said Charlie viciously.

"What do you say, Brown?" inquired Morton.
"Shall we let Charlie go?"

"How do you propose getting down the hole?"
asked Brown.

"We can climb down," returned Charlie.

"I don't think you'll find either Billy or
Columbus go far with you," said Morton.   "If
we had any sort of a rope I should not mind,

but there's nothing but the tent-line, and that's not strong enough."

" I'll take great care," pleaded Charlie.

" No doubt you will, but if you make a slip and flop down some thirty or forty feet, no amount of care will get you back again with sound bones."

Charlie looked unconvinced.

" We could keep a look-out for the absent natives," said Brown. " They are bound to come along this track, I suppose."

" Could not Columbus make some sort of a mark to stop them?"

" I'm afraid not. Blacks can communicate in some way—you have seen their 'yabber-sticks', I suppose,—but I don't think we could make Columbus understand what we wanted, nor do I suppose he would do it if we could."

" Strange how they can communicate, though. The time Faithful's party were murdered in Victoria, the blacks in the settled districts knew of it long before the whites did."

" Well, can I go or not?" demanded Charlie.

" I'll sleep on it," replied Morton. " I think you can take care of yourself, and I can trust Billy as far as one can trust a blackfellow. But remember, I am responsible for you, and if anything happened to you I should be to blame."

With this Charlie had to content himself until the next morning.

Morton and Brown stopped up late, smoking by the fire.

" Shall you let the boy go?" asked Brown.

" I think so, but I'm doubtful of those fellows behind; they might slip past us in the dark and fall foul of Charlie when he was not expecting them. If they had fair play Charlie and Billy could hold their own, but they might take them at a disadvantage in amongst those boulders."

At this moment a wailing cry in the distance made them both start. The cry exactly resembled the mourning lamentations they had heard in the scrub.

" That settles one part of the question," said Brown. " Those fellows are on their way back. Kick old Columbus up and get him to answer them."

Morton promptly roused the slumbering chieftain, and when he heard the approaching cry he at once answered it. Then he went out to meet them. Apparently he soon told them all about the catastrophe that had taken place, for presently a great cry went up. Columbus soon after appeared, leading them into the firelight. Six truculent-looking ruffians they were, but it was evident that Columbus had impressed them with a due respect for the power of the whites, to whose anger he attributed the misfortune that had befallen their tribe, for they all wore a very humble and downcast air.

Charlie, who had come out of the tent on hearing the noise, gave them some food, and they made a fire apart and squatted beside it, Columbus being cautioned against allowing them to sneak off during the night. As the blacks were unarmed, and they now had all the survivors under their direct observation, no watch was kept, and the late enemies soon slept soundly without any misgivings.

CHAPTER X.

Stuart's Journal—News of the Missing Expedition—
Charlie Departs.

NEXT morning Morton told Charlie that, as the natives had turned up, he could go and explore the site of the crater, but he must be back within three days. Columbus was made to understand that the six blacks were to remain in the camp, otherwise they would share the fate of their countrymen. As there was a good supply of game, fish, ducks, and pigeons, they could easily live without trespassing on the rations of the whites. As Charlie was not to leave before an hour or so before sundown, he had ample time to make his preparations. Meantime the others went seriously to work transcribing the journal, which took them the greater part of the day.

The result of their labours was as follows:—

Stuart's Journal.

"September, 1848.—We have been fortunate in striking a well-watered river, but—" (Here there was a portion mutilated.)

"September 12.—Still these aimless journeys to the westward, across plains, barren and waterless, and which are so loose and cracked that every time we make an attempt we lose some of our animals. Fortunately, we have this fine river running north and south to fall back on. The men are very discontented, and the prospect ahead is anything but bright.

"September 16.—Two more horses knocked up by this obstinate pushing into an impassable desert. Klausen and I must remonstrate seriously to-night.

"September 18.—Thank Heaven, we managed to make the Doctor see his folly, and we are now on the move north to get on to his old track and work round that way. Everything going on much smoother.

"September 20.—Country still well watered, and travelling easy; expect soon to—" (Here the journal was undecipherable to the end of the page, and the succeeding one was dated more than a month ahead.)

"November 2.—We are still on the Doctor's first track round the foot of the Gulf; but although there is ample feed and water our horses are falling away, and do not look as well as they did in the dry country.

"November 3.—Had some trouble with the natives yesterday when crossing a small river. We had to fire twice at them before they ran away. Klausen was speared in the arm with a barbed spear, which had to be cut out."

(Here there was another gap where the journal had been mutilated.)

"December 15.—We are still camped on the river, which the Doctor called 'The Roper' on his first trip. Klausen's arm prevents our moving. Inflammation has set in and I am afraid he will die. Blacks very troublesome.

"December 16.—Klausen died last night; we buried him this morning. We now leave the Doctor's old Port-Essington track and follow this river up, south and west.

"December 20.—Getting into dry country again, and the scrub is becoming very bad. We are scarcely able to force a way through on foot—"

(Here, for many pages, the journal was so mutilated and discoloured by water that only an occasional line was intelligible. These seemed to point to the party being constantly baffled by scrub and dry country; and also that some of them were attacked by scurvy.)

"You've been in Queensland, Brown?" said Morton when they arrived at this stage in their transcription. "How do you follow out this journal?"

"As plain as print. Stuart's journal—at least the part we have—commences on what is now known as the Diamantina River, I think. The great dry plains he speaks of are to the westward of that river, and in a dry season would be impassable to anyone not knowing the country. By following the river up they would easily cross the watershed on to the Gulf of Carpentaria waters, and so get on to Leichhardt's old track."

Brown got a map out of the pack and illustrated thereon what he had just said.

The next coherent portion of the journal would seem to have been written after a disaster.

"April 24th, 1849.—There are now only five of us left; two—Hentig and the Doctor—are both sick. The other two must have died on the dry stage, as they have not come in here and the blacks would not let them go back. I have not been able to write my journal for some days, and as the Doctor cannot write now, no record at all has been kept. We were just packing up to leave the rocky water-hole in the scrub where we had been camped for some days, when the blacks attacked us on all sides. There were so many of them, and they had such good cover in the scrub, that we fairly had to get away as best

wo could without water, or all of our packs.
While we were trying to keep them off a gun
burst and nearly shattered the Doctor's hand.
This forced us to hasten our retreat to get him
safely away, leaving some of our horses and
mules behind. Immediately on getting out of
the scrub we found ourselves in open stunted
forest, covered with prickly grass. We kept to
the south-west until evening and then camped
for a while, for the Doctor and Hentig could go
no further. We had travelled very slowly, and
when we camped Kelly asked me how far I
thought we had come. I told him about ten
miles. He then proposed that we should take
the freshest horses and go back and try and get
some water, as even if the blacks were camped
at the hole they would be asleep at the time we
got there. I agreed with him, for it seemed
hopeless to go on through this dry forest without
water. I suggested, however, that we should
take all the animals and Murphy, and if possible
give the blacks a fright. Leaving the two others
to look after the sick men and keep a fire going,
we started, and were singularly fortunate. We
got back soon after midnight and found the
blacks camped by the water-hole. They were
asleep, having been feasting on the horses we were
forced to abandon. Some awoke, however, and
we immediately rushed into the camp, shouting
and firing. They fled indiscriminately, leaving

most of their weapons behind, and these we
heaped on the fires. We were lucky enough to
find two big kegs we had abandoned, and filling
these and all the canteens we had brought, we
started back as soon as possible before the natives
recovered from their scare. We reached camp
soon after sunrise, and but for the success of
our raid none of us would be alive now, for that
dry forest continued without change or break,
day after day. We hoarded up some of the
water for the sick men and managed to keep
them on their horses, but I remember nothing of
the last day, nor how the other two parted from
us. Murphy says they went after what they
insisted was a smoke, but he says it was only
a whirlwind passing over burnt country. Kelly
found this water-hole through seeing two white
parrots coming from this direction early in the
morning. It is on the edge of the forest, and to
the west lies a great plain, still covered with the
same prickly grass. There is a little coarse grass
for the few beasts we have now left, but the
water in the hole is thick and muddy and fast
drying up.

"April 25th.—We have been back to try and
find the other two men, but without success; we
must stay here—"

(Another break in the narrative here came in,
the paper seemingly having been scorched by
fire.)

As it was now getting on for the hour when Charlie's departure on his trip was to take place, the two men knocked off their work and assisted him to get away. Fortunately, having Columbus with them, they were enabled to lighten the packs considerably, as they made him carry his share. Morton parted with his young relation with some misgiving—still he liked his pluck, and did not care to baulk him. By the time the sun disappeared the three of them were mere specks in the distance of the great plain.

The six natives seemed quite contented to stay where they were, but both Morton and Brown determined to keep a sharp eye on them. If they discovered them trying to make for the rock-plain they could easily overtake them on horseback before they could cross. However, they were there in the morning, and Brown and Morton settled down to the continuation of the journal.

CHAPTER XI.

Continuation of Stuart's Journal.

THE narrative now assumed a more connected form, telling of the death of Dr. Leichhardt and the rescue of the three survivors by the friendly natives; also of the discovery by Stuart of some curious cave paintings, which bore evidence of being the work of a race superior to the present inhabitants of the interior.

Continuation of Stuart's Journal.

"——Ever since the Doctor injured his hand through the musket bursting he has been subject to attacks of feverishness and temporary madness, and this has greatly added to the hopelessness of our position. I have often asked him for some definite statement of his intentions, but he seems quite unable to go into any details, and I am afraid we are fearfully out in our reckoning. Hentig still terribly bad with scurvy.

"May 1, 1849.—Since my last entry we have buried Hentig, and the Doctor must soon follow. If we could only get across this dry country ahead of us we might be able to move on, but since we are almost without rations and most of our horses dead it seems as though we must leave our bones here, for there is no turning back. Doctor much worse. Kelly says that there is only two days' more water left in the hole. No sign of rain. Weather getting cooler.

"May 2.—This morning, before the sun got up, I climbed the tall tree on the edge of the plain, and distinctly saw a faint smoke to the westward, in the same direction that Kelly thought he noticed it when we first came here. Tomorrow we will start towards it; it is all we

THE DEATH OF DR. LEICHHARDT IN THE DESERT.

M 64

can do. How we shall get the Doctor on I cannot tell; he is almost helpless, and his mind is quite gone. We have four horses and two mules. Besides the Doctor, whom I look upon as a dead man, there are Kelly, Murphy, and myself (Stuart). Hentig is buried under the tree with the cross cut on it. Klausen died on the river Roper five months ago. I will bury a copy of this in a powder-flask in Hentig's grave, as well as the Doctor's papers."

(Here there was an evident gap in the narration.)

"I have been too ill to write for many days, how long I don't know, for we have all of us lost count. I am only just beginning to remember our journey across that horrible desert. We started in the direction I had seen smoke, after using up every drop of water for the animals at the camp where Hentig is buried. We took it in turns to hold the Doctor on his horse, but he got very bad a few hours after we started, and when the sun grew hot he begged us to lift him off the horse for a little while. We had all the canteens full, and Kelly had made a bag of calico and rubbed it outside with goat's fat, and it held water tolerably well. So we gave the Doctor plenty to drink, but he got no better, and about noon he died. He talked a good deal to himself in German, but had lost all knowledge of us or where he was, and a good thing too. We could not stop to bury him, for we had to push on, so we left him there on the big plain, where I think no living thing ever comes or ever will come since we were there. It was the second day out when we got on to that prickly grass plain with deep red sand, and then our horses began to give up, and we had to walk and try and drive all the beasts; but they were so thirsty they would not keep together, so we stopped and talked about what was best to be done. Kelly and I agreed that it was best to unpack all the animals, and, taking all the water and as much food as we could carry, to march on, and perhaps if we soon came to water we could come back for the beasts. Murphy did not think so; he thought we could drive them on on foot when it got dark, but we persuaded him that we were right, and started. We walked on long after it got dark, and then we lay down and slept on the sandy plain amongst the prickly grass, for since leaving the camp we have never passed a tree. In the night Kelly called me and pointed to a light in the west, which was evidently the reflection of a large fire. Next morning we met it, for a wind had sprung up in the morning. It was well for us that it was almost barren country where we met the flames, otherwise we would have been burnt. As it was we were nearly stifled with heat and smoke. Afterwards, all that day and night, we watched the glare of it behind us blazing amongst the dry prickly grass we had passed through, carried on by the strong wind that now blew, and we knew that our animals, saddles, and the Doctor's body would be burnt up, and no one would ever see more of them; so it was with sorrowful hearts we walked on. That day I saw some trees on ahead, and we turned—"

(Here the journal had been effaced, apparently by water, but nothing of importance appeared lost.)

"Murphy was the weakest, but we stood by him, although the burnt country was very distressing. Kelly got a little light-headed towards morning, and I began to feel the same. I don't remember much more; it all seems a dream of stumbling along and helping each other, sometimes talking to the phantoms we all fancied we saw walking with us; and then I came partly to my senses under a rough shade of boughs, and before me was this great lake, and I knew by the smell of the place and almost without looking around that I was in a camp of the natives. Kelly and Murphy were alive, and better than I was. They remembered something about the natives helping us to the water, for we had passed it, and were going right away—"

(Another gap.)

"June.—I have put down June, for I think it must be that time of the year, as near as I can make it. Neither Kelly nor Murphy can read or write, so while I was ill they did not keep the dates. The natives are quite friendly, and Kelly, who was born in the bush of the southern portion of the colony, has attained great influence over them, as he is very active, and can use nearly all their weapons. I have been round the lake; it is nearly sixty miles round, but very shallow, except at the end where our camp is. The natives tell me it dries up some seasons, with the exception of the deep hole here. They have canoes made out of shells of trees, and can manage them very well, standing upright and poling or sculling with a spear. They know nothing of any other blacks, excepting a tribe to the eastward, of whom they seem greatly frightened. They are a very simple people, and live well, as there is plenty of fish in the lake and wild fowl. Kelly and Murphy have quite settled down to the life, but how different it is for me! When I think of my own people, and how I am doomed to live and die amongst savages, I nearly go mad; for unless other white men find their way here I must die here; and who would cross that horrible sand plain? If the Doctor had but lived we might have found some way of escape, but he and our horses and saddles are all burnt up. What is the good of keeping this record? No one will ever read it. I will become a savage like those around me, and forget what I was.

"July.—I must write or I shall forget my

language, and that I must keep while life lasts. A strange thing happened to-day. The old man Powlbarri came to me and made me understand that he wanted me to go with him, he had something to show me. I followed him to the ridge where the great sandstone rocks are, and he led through a gap between two of them so narrow that we could scarcely squeeze along. In a short time we stood in a spacious cave that penetrated seemingly into the depths of the ridge. There was a bar of limestone in the side and a few stalactites, but not many, and light was admitted dimly through cracks and crevices overhead. But when my eyes were more accustomed to the light I started with affright, for partly overhead and partly confronting me was a strange gigantic shape with outstretched hand. I recoiled for an instant, and then saw how I had been deceived; it was a rock painting on the sloping roof of the cave. It bore no resemblance to the ordinary crude tracings of the natives. I looked at it narrowly, and tried to get out of the old man who did it. He gave me to understand that it had been always there—as well as I could comprehend him,—longer than the blacks knew of. The figure was of heroic size, with straight symmetrical features, the head surrounded by a halo or turban, and the body attired in a rough semblance of a robe. The whole figure was of grave aspect, and much reminded me of the drawings I had seen of Egyptian gods. The old man beckoned to me to withdraw, and I was not sorry to do so, for I wanted time to think, and intended to come back with Kelly and Murphy and explore the place thoroughly. We passed out of the cave, and had just squeezed through the gap when our ears were greeted with a shrill discordant yell of terror from the camp. With an answering shout my companion with extraordinary agility bounded from my side, and I ran after him. There was little doubt what had happened; the dreaded tribe from the east had surprised and attacked the camp. When I arrived on the scene the fight was just assuming extensive proportions. At first the boys, gins, and old men had been easily overpowered, some killed and some captured; but a hunting-party came up, amongst whom were my two companions, who now went naked, and were nearly as dark as the natives. Kelly, who would have made a brave and dashing soldier if fate had so willed, plunged at once into the thick of the fray, followed by Murphy, who was slower in his movements. There appearance disconcerted the enemy, who were horribly distinguished by a red smear on the forehead and a white triangle on the breast. They rallied, however, but Kelly's onset, so different from the ordinary method of native warfare, had evidently staggered them. I was about to join our side when I remembered that nearly the only part of our equipment saved was my double - barrelled pistol and ammunition bag.

This I had never used, reserving it for our own protection, and I ran to my whirlie and came back with it loaded. A tall blackfellow was engaged with Kelly, and rushing up I fired at and shot him. There was an instantaneous lull of surprise, and at the discharge of the other barrel the attackers straightway fled, and even our own side seemed inclined to follow their example. Alas, our victory was dearly bought! Kelly was speared through the chest, mortally I saw at once; and so it turned out, for the poor boy died with his head on my knee in a few minutes. We buried him that evening, and never did I feel more sorry than for my bright young companion, who, although uneducated, had many noble qualities, and—"

(Here there was a large portion of the journal quite undecipherable; the few words distinguishable seem to point to a visit to the cave with Murphy.)

"The strange mystery of the cave paintings still puzzles me. The additional smaller drawings we discovered are most singular, and certainly point to other authorship than that of the natives. In many places there are signs like a written language, and the peculiar portrayal of dress indicates an Asiatic origin."

(Another gap.)

"I miss Kelly still. Murphy is dull and intractable; he has sunk to the level of his savage companions. O God, have pity on me, for I shall never never see my countrymen again! Surrounded by deserts, impassable to me on foot, I must drag out my life here, hoping for the succour that will come too late to save me."

(Here the narrative broke off, although several more blank leaves in the pocket-book were available.)

CHAPTER XII.

Charlie's Adventure.

"WELL," said Brown, "which is it to be? South or west?"

"According to Columbus, Stuart was down with the southern tribe the last time they saw him, which is apparently many years ago."

"And he says that the road to the southern tribe is the easier to travel. I think we ought to go there first."

"Then tackle the western lot. We must thoroughly examine those caves he speaks of."

"Yes, the horses are in fine hard condition now; we will make a start as soon as Charlie comes back."

"We ought to go round by Hentig's grave and recover those papers."

"We have got our work cut out. Lucky we brought a good supply of rations."

The six niggers appeared to have settled down

contentedly to await the return of Columbus. They were not at all intelligent, and both men failed in getting any further information from them.

"What's to become of these beggars when we leave?" said Brown. "We must take Columbus with us to show us the best road."

"There's plenty of game here, and up and down the water-course they will be able to earn an honest living," returned Morton. "There's not enough of them to resort to their cannibal practices again."

"I sha'n't be sorry when Charlie comes back; I am tired of doing nothing."

The time appointed for Charlie's return drew near without any sign of the three men. Morton watched the plain all day, finding it impossible to conceal his anxiety, and blaming himself for having allowed the boy to go.

At last, not long before sundown, a solitary figure was seen approaching. Morton eagerly snatched up the glasses.

"Columbus. And alone," said he, putting the glasses down with a sigh.

The two friends waited anxiously for the approach of the native. Instinctively they felt that some disaster must have happened.

As soon as Columbus was within hearing he commenced howling dismally, and the six others answered him, lamenting in a loud voice. This was kept up at intervals until Columbus reached the camp. Without waiting to be questioned, he held up two fingers and pointed down to the ground. Charlie and Billy were evidently in trouble somewhere underground. Brown indicated that they would go out there, which was evidently what Columbus wanted.

"We must take all the surcingles," said Brown; "they will bear Charlie's weight, I think, and will make a good long rope buckled together."

"Columbus has been evidently sent back to bring assistance; and the old beggar has travelled too, by the look of him. What do you say to taking two of the others with us?"

"I suppose they will not sack the camp while we are away?" returned Brown.

"No, they are not civilized enough for that. Now let us make all the haste we can."

Columbus was instructed to tell two of the blacks to accompany them, and to explain to these men what had happened. This he did in several rapid sentences, and in a few minutes they were ready for the road. Their equipment was but light, as they only took their revolvers, candles, the surcingles, and a little food and some brandy, a small supply of which they had with them; this with their water-bags would be all they required, they reckoned. They pushed on with scarcely a rest all night, and found the advantage of having the natives with them, for they could not have found their way amongst the boulders in the dark.

About an hour before daylight they stood once more on the edge of the hole in which the crater had sunk. There was a decided bad odour arising from it, distinctly noticeable at that time.

Morton leaned over the edge and shouted "Below there!" as loud as he could. There was silence for a second or two, and then "Below there!" came thundering back.

"Echo," said Brown.

Morton tried again with the same result.

Brown fired his pistol, but the thunder of the echoes was the only answer.

"They must be poisoned with foul air," said Morton, in tones of the deepest sorrow.

"Must we wait until daylight?" asked Brown.

"I am afraid so. We might come to grief ourselves, and then it would be all up indeed. However, I think I can get down to the edge of the fissure without much danger, if you and the two blacks can hang on to the surcingles."

The preparations were soon completed, and Morton carefully made his way down the sloping sides of the hole and amongst the mud-encrusted boulders, by the help of the surcingles, which Brown and the two natives held above. It was slow work, for the candle he had gave out only a feeble light, but at last he found himself at the edge of the rift at the bottom. He stood there listening for some time; presently, with an up-blast of cold air that nearly extinguished his candle, came a strange wail as though some giant was sighing, far underground.

"Hear anything up there, Brown?" he shouted.

"Not a sound. Are you on level ground, can we slack off?"

"Yes, slack off. But do you think you could trust the two blacks to hold it while you come down? I will come back and show you a light."

"I'll chance it at any rate," returned Brown, and presently he stood beside his friend.

Morton told him of the strange sound he had heard, and both stood by the edge of the hole and listened. Once more the blast of cold air came and with it the melancholy and mysterious noise.

"That's no human or animal noise," said Brown; "it seems more like water or air escaping."

"The atmosphere does not seem so bad now," said Morton. "I suppose it was the contrast with the pure air above."

"It was getting light to the eastward when I came down just now," returned Brown; "we had better wait for full daylight—half an hour cannot make much difference."

"It might make all the difference," replied Morton; "however, I suppose there is no help for it."

At that moment there was a sudden cry from above.

"Wonder what's up?" said Morton, scrambling

back. "Hang it all!" he exclaimed as he laid hold of the surcingles and they came tumbling down, showing that the blacks above had let go of them.

Presently they were heard jabbering at the edge of the hole, and Morton shouted to them and threw a coil of the surcingles up. Apparently they understood what was wanted, for the line tautened once more and Morton scrambled up, and then assisted Brown. The dawn was rapidly breaking, and the blacks, pointing to the candle Morton still held in his hand and then towards the memorable cliffs, chattered volubly.

"They must mean that they saw a light in that direction," said Brown "It's too light for us to see now."

"Shall we go over there or investigate this hole?"

"They must have seen something by the start they got; perhaps we had better go there first."

Accompanied by the two natives, who led the way by a path known to them, they made for the shattered cliff which they had hoped never to see again. As they approached it an awful odour, evidently stealing through the cracks from the bodies rotting beneath the collapsed roof, made itself disagreeably evident. The blacks kept on talking to one another, as though discussing what they now saw for the first time. Arrived at the place, the white men mounted on the piled-up débris, and both together shouted with the full strength of their lungs. To their delight a distinct answer was heard from beneath their feet, evidently no echo.

"It's Charlie's voice!" cried Morton delightedly. "Where are you?" he yelled.

"Here!" came the voice, right under their feet.

"In the old cave?" asked Morton.

"No, but close to it; there's a vile smell here."

"How did you get there?"

"Came underground. Do you think you can get us out here, or must we go back again to the crater?"

"The blacks saw your light, so there must be a big gap somewhere. Have a good look round, it's sunrise now."

There was a silence for a while, then Charlie's voice sounded a little further off.

"There's a big crack here, but too narrow for us to get through."

The men above went in the direction of the sound, and soon found the fissure.

"It would take dynamite to shift this one," said Brown, putting his hand on the huge boulder that formed one side of the rift.

Morton knelt down on the flat rock that formed the opposite side and put his hand into the crevice.

"Hurrah!" he shouted, "this is only a slab; the four of us can shift this, I think."

So it proved. The four, with one unanimous

pull, managed to partly upraise the limestone slab, and Morton adroitly kicked a big stone underneath which kept it in position. Charlie was now able to crawl out, followed by Billy.

"Here, take a swig of this before you say anything," said Morton, mixing some brandy and water in his pannikin.

Charlie took some and handed the remainder to Billy, who looked particularly scared.

"How long have you been cruising about in the bowels of the earth?" asked Brown.

"Since the night before last, or early yesterday morning. I did not go down the deep hole when I first came here, because I had promised to be careful. I went down the first one, and then got some dead leaves from the old bottle-tree camp, lit them, and threw them down. By the light we could see that there were no sides to the hole—it seemed as though it had been punched in the roof of a tunnel. However, we found a place at last where some boulders were piled up, and I thought that, with Billy's help, I could get down on to them. I did so, and found that I could get from there on to the floor of the place without any trouble. I came back for Billy, and he was being helped down by Columbus, when suddenly there came a most awful sound, half a shriek and half a sigh, which so frightened Billy that he must have let go, for he came tumbling on top of me, and the two of us dislodged the boulder, which was not very firm to start with. Fortunately we were neither of us hurt, but Columbus must have thought we were killed, for he cleared out—"

"And came straight back to us, luckily," interrupted Morton.

"When I found that he was gone," went on Charlie, "I thought we could get back again by piling rocks up to stand upon; but there were no small ones, they were all too big for us to shift. We waited there, and shouted, and called, and every now and again we heard that sigh—"

"We heard it as well," said Morton.

"Billy shook with fright every time, and nearly made me as bad as himself. At last I made up my mind to explore the passage we were in; but I had a great job with Billy, for the passage led in the direction the sound came from, and Billy conjured up all manner of horrors. Luckily I had the packet of candles with me when I came down, so we had plenty of light."

"Was the air bad?" asked Morton.

"There was a funny damp smell, but the candles burnt well, and we felt no bad effects. The passage was smooth enough underfoot but not very high, so that we had to stoop; but we came to occasional places where we could straighten our backs. The noise kept getting louder, until at last Billy with his terror got to be such a nuisance, that when we got to where there was enough space I put my candle down and gave him a good punching."

Here Brown and Morton burst out laughing. The idea of Charlie, down in the blackness of a subterranean passage, thumping Billy to keep his own courage up, was too original.

"Presently we came to where the passage branched, and along one came the noise, now a regular bellow. Nothing could induce Billy to go along that one; he threw himself on the ground and let me kick him, but he wouldn't budge."

"Were you very anxious to go yourself, Charlie?" asked Brown.

"I had to keep up appearances," returned Charlie modestly. "We started along the other passage, and presently it began to ascend, and was littered and partly blocked with boulders; finally, after much trouble and squeezing, we got up to where you found us. It was dark when we got there, but I knew by the fresh air coming in that there were some cracks somewhere leading to the upper world, and I guessed by the smell that we must be in the neighbourhood of the old cave. We went back a bit and lay down to sleep, and when I woke up we came here again so as to be ready to try and get out at the first dawn."

"Thank goodness it's all over," said Morton; "for I've had a rare fright, and Brown and I have been travelling all night. However, we won't go back without investigating the mystery of the noise."

There was still some water lying about in the rock holes around the crater, so when they returned they set to work and got breakfast ready.

Charlie thought with them that the strange noise was made by an escape of water or air, both from the regularity of the sound and its peculiar nature.

CHAPTER XIII.

Investigation of the Mysterious Noise—The Trip South— Natives Exterminated—Stuart's Initials Found.

AS soon as the meal was finished, Morton, Brown, and Charlie descended the hole, but Billy declined the invitation extended to him. By the aid of the surcingles they climbed down into the passage already traversed by Charlie and Billy, and which appeared to have been an underground water-course at the time when the boiling springs were at work.

At last they arrived at the branch passage from whence came the mysterious noise, and along this they proceeded cautiously. Suddenly across their path extended a black chasm, bringing them to a stand-still. Testing the ground carefully, they crawled to the edge and looked over. The darkness was intense, but on holding the candle out, a tiny spark was reflected down below, as though from the surface of a sheet of water. Suddenly this disappeared, and the loud sigh, or, as Charlie had called it, "a regular bellow", came up from the pit. This died down, and they heard a repeated swishing noise, like water splashing against rocks. Morton inverted his candle so as to get the wick well alight, and then dropped it down the hole. They watched it falling for some time ere it struck the water with a distinctly audible hiss.

"By Jove! it's a long way down there. I don't think we need stop here any longer unless Charlie wants to go down."

"No, thanks," said Charlie, "my curiosity is quite satisfied."

They retreated as cautiously as they had advanced, followed by the melancholy roar from below.

"It's the water that makes that noise," said Brown. "The water down there is evidently still in a disturbed state, and is regularly set in motion, and rushes up some sort of a blow-hole."

"Do you think it has any connection with that hot swamp and lake down south?"

"Without any doubt; perhaps we shall find that lake all burst up when we go down there."

They retraced their steps, and, by the aid of the surcingles held by the blacks above, emerged once more into the open air. They rested most of the day, and started back to what they now considered their main camp as soon as the evening grew cool.

Columbus and the other four blacks were there, and everything as they had left it.

"That was a smart trick," said Brown.

"What was?"

"In our hurry to start after Charlie, we left the journal and the copies we had made lying loose in the tent. If the grass had taken fire, or the niggers looted the camp, we should have lost all our work."

Morton whistled.

They rested one day, and then made an early start for the south. They had rigged up a make-shift saddle for Columbus, and, as they travelled slowly, he was able to get along fairly well. They reached the swamp about the same time as before, and at first noticed no change in it. On penetrating it, however, they found that it was not so boggy as formerly, and on mounting the tree they had already climbed, they saw that the water in the lake had fallen considerably, and the fringe of reeds was drooping.

"Do you think all these fine trees will die if the water dries up?" said Charlie.

"They may. But as their roots go down to a great depth, I should think they would hold on a good many years yet."

Next morning Columbus indicated the track they had come upon before, and they soon had left the swamp behind them. The country was exceedingly monotonous, there being no break in the forest until about four in the afternoon;

then they suddenly came to a creek, and the country began to improve, and better grass was apparent. At the first water they came to they camped for the night. Columbus intimated to them that the creek they were on and the one where they had been camped were the same, and as the characteristics were similar, they concluded that he was right, and that the creek had re-formed again. Columbus also informed them that they would now follow the creek, and that there was plenty of water all the way. On inquiry, he said that they would reach the mountain in two days.

On the evening of the second day they got into broken country, although it was still well grassed, and the creek had largely increased in size. From the crest of one ridge they passed over, Columbus pointed to the mountain now visible in the distance.

Next day the country was much rougher, and the creek ran through a succession of gorges. The mountain was the highest point of the broken country, and the creek swept round the base of it.

Morton called Brown's attention to the fact that all the native camps they passed were of old date, and that no fresh tracks were visible. At last they reached an extent of open country lying at the foot of the mountain, which rose aloft in a peak. On the bank of the creek were some ruined humpies, built of mud and sticks, after the manner of the Cooper's Creek natives. In the creek was a long water-hole, apparently of great depth. Human bones and skeletons were strewn about the camp. Evidently a wholesale massacre had taken place some years back.

"Those cannibals must have wiped the whole tribe out the last time they were here," said Brown.

"If they have served the tribe to the westward the same way, they would have had to live on one another shortly."

After unpacking and hobbling the horses, they made a thorough investigation of the place to see if any trace of Stuart still existed, but they saw nothing to lead them to suppose that a white man had ever been there.

Columbus, on being appealed to, pointed to the hill, which was scarcely a quarter of a mile away. On going over to it they found what appeared to be a crude kind of barricade built of stones, a work that none of the party had ever before seen done by natives. This was the only indication they found that evening. The next morning early they ascended the hill, and from the top had an extended view all around. They were evidently on the highest point in that part of the continent. To the south and east it appeared to be one vast ocean of scrub, without a break to the horizon. They could trace the course of the creek for some short distance, then

it apparently died out and was once more lost.
Westward the scrub was broken into belts and
patches, until it merged into a wide gray plain, to
which they could see no end but the sky-line.
Northward was the broken country they had
passed over. The mountain was of granite formation, and on a smooth boulder they found
some initials plainly chipped on the surface:
"C. N. S. 1861".

"Stuart was here, then, right enough. I wonder whether he went on from here."

"He never got into the settled districts at any
rate, or we should have heard of it."

Columbus, who had accompanied them, shook
his head when asked about the country to the
south and east. He made a gesture like a man
falling down dead, by which they understood
that it was impassable, so that the probability
was that Stuart had perished in his attempt to
make into civilization.

Brown struck a match and lit his pipe.

"We have come to the end of our tether in
this direction," he said.

"I wonder how the lake bears from here,"
replied Morton. "I suppose the cannibals have
a track from the great rock out to it, but if
Stuart got down here on foot we ought to be
able to find our way across on horseback."

Columbus, on being questioned as to the direction of the lake, pointed north, the way they had
come.

"That's a way the niggers have," said Brown. "They always point to the last place they started from; they have no idea of direction. When we got back to the old camp he would point some other way."

Columbus professed entire ignorance as to any means of reaching the lake, except by going back the way they came and starting on the road he knew. Morton and Brown, however, decided on trying to go straight across from where they were.

They devoted one day to a trip down the creek, which they found was entirely lost in sandy, scrubby country. No further sign of Stuart's presence was found anywhere, nor could anything be discovered to lead one to suppose that any of the natives had survived the massacre, although Columbus had evidently expected to find some still living.

Calculating the supposed situation of the lake as due west from the rock, they reckoned it would be north-west from where they then were. If that course did not bring them to the lake they would probably come across some indication which would lead them to it.

The first part of their journey was through the belts of scrub they had seen from the hill-top. It was principally hedgewood, and greatly delayed their progress, and it was late when they at last emerged upon the edge of the plain. The grass was fairly good, but there was no water for the horses, and from what they had seen there did not seem much prospect of getting any early the next day. In fact, it was past noon before they had crossed the plain and gained the timber on the other side of it. This was open forest, and in a clear space, some mile or two on, they came to a dry lagoon. In the shallow bed was an old native well, and on clearing this out and deepening it, a very fair supply of water came in. Watering their now thirsty horses took some time, as all the water had to be drawn from the well, a billyful at a time, and poured into a trough extemporized from a waterproof sheet. The supply, however, came in strongly, showing there was a good permanent soakage. There was fine feed about the lagoon, and everybody felt satisfied with the prospect ahead.

Columbus seemingly knew nothing of the country they were then on, so that the cannibals had evidently stuck to one particular track when on their periodical man-hunting expeditions.

CHAPTER XIV.

In the Spinifex Desert—Arrival at the Lake—The Remnant of a Tribe.

THE next morning when they started the forest country still continued for many miles, until they at length came to another broad plain, and a couple of hours before sundown sighted some timber nearly on their course. This proved to be a double line of gutta-percha-trees, with a broad flat between them. The trees grew on low banks of sand, on which were countless quantities of tiny shells; the whole had the appearance of a shallow water-course, but the bed was covered with blue-bush. The two lines of trees stretched on like a limitless avenue, and as it followed much the same course as that they were travelling, they proceeded along it. They passed one or two empty holes, with a ring of polygonum bushes, dry and withered, around the top of the bank. It grew late, and as everything still bore a parched appearance Morton pulled up for a consultation.

While discussing the best thing to do, a flight of galar and corrella parrots passed overhead, flying in the direction they were going, and evidently making for their nightly drink. This put new life into everybody, and they pushed on once more. At dusk they were rewarded by coming to a somewhat deeper hole than those they had passed. There was, however, only sufficient water for their wants in the bottom, and it was fast drying up, and could not be depended on for their return journey.

Next morning they still kept on along the avenue of gutta-percha-trees, and Morton began to hope that it would turn out to be one of the water-courses supplying the lake. In this, however, he was disappointed, for the trees grew fewer in number and further apart, until they passed the last one, and before them stretched once more a boundless plain.

The country now suddenly changed for the worse; the ground was sandy and covered with the detestable spinifex, and both Morton and Brown felt rather doubtful as to going on, for there was no knowing how far this desert might extend.

However, they made up their minds to proceed, as there was really nothing else to do. Then commenced one of the weariest rides they had yet experienced during the trip. It was even worse than the scrub. The prickly needles of the spinifex irritated the shins of the horses, so that it was with great trouble the pack-horses were urged along. Hour after hour went on, and still there was no change in the unbroken horizon that bounded them.

"I should fancy those old cannibals found it

mighty rough on their shins if they had to cross
a belt of desert like this," said Brown.

"I expect they kept it burnt down on the
track they used to patronize," replied his friend.

"Fancy what the feelings of poor hopeless
Stuart and his companions must have been when
toiling through this waste."

"Yes. If we find it bad, what must starving
men on foot have found it?"

That night fell on them still in the desert.
They had an ample supply of water for themselves in the canvas bags, but their horses had
to go both hungry and thirsty.

"Things begin to look rather queer," said
Morton, as they prepared to start.

"Yes, it's a case if we don't get out of this by
to-night."

They had hardly mounted, when Billy and
Columbus gave a mutual exclamation, and pointed
to the westward of their track. A curious looking dark mass was travelling swiftly along just
above the horizon. Suddenly it dipped down
and disappeared.

"Hurrah!" shouted Brown. "Flock pigeons
going in for their morning drink. That must be
the lake."

Much elated, they pressed eagerly on in the
new direction, the horses seeming to understand
what was ahead as well as their masters. The
spinifex now began to grow scantier, and patches
of grass appeared in its place; the earth changed

from red sand to good chocolate soil, and before them stretched a very large expanse of downs, well grassed with Mitchell grass and other good grasses.

Suddenly and unexpectedly they crossed the crest of an imperceptible rise, and before them lay the goal of their hopes. Unanimously they halted and gazed at the locality where the man, whose journal they had read, had passed many weary years of exile.

No fairer scene could have been found anywhere in the interior of Australia. A blue expanse of water, apparently some miles in length, lay outstretched before them. The low sloping banks were verdant with grass, kept green by the soakage from the lake. Great gnarled coolibah-trees of immense girth grew round the water's edge, and the gently rising downs on either side were studded with clumps of the beautiful weeping myall and shady bauhinia trees. At the end of the lake nearest to them was a small hill crowned with great gray boulders of granite.

"After all, there must be something in the influence of surroundings," said Morton. "The natives living here, are, or were, according to Stuart, a gentle and friendly tribe, whilst those living amongst the barren rocks alongside of that boiling spring were about the biggest devils I ever met."

"How about Columbus?" asked Charlie. "If

there are any natives left, won't they try and kill him?"

"No doubt they will make it pretty sultry for him, but he seems quite cheerful over it. He has put the onus of the whole thing on our shoulders. He must take his chance."

They rode on in suppressed excitement, hoping against hope that Stuart might still be there.

"What's that ahead?" suddenly cried Morton.

The object when approached turned out to be the bald, dried, half-decomposed body of a blackfellow.

"This is some of your men's doings, my friend," said Brown, glancing at Columbus, who grinned complacently.

The sides of the lake were firm and hard, and the thirsty horses ran eagerly in and commenced drinking greedily. Overhead the white correllas and pink and gray galars chattered noisily amongst the boughs, on the opposite side a group of objects like native gunyahs were visible. When the horses finished drinking, they rode round the edge of the lake to them.

Not a sound was heard, and nothing was seen to move as they approached the spot, nor was any smoke visible. Gorged carrion crows and hawks arose as they drew near, and flapped unwillingly away; the crows protesting loudly at being disturbed, after the manner of their kind. Two or three eagle-hawks gazed fiercely down from the branches of neighbouring trees.

A pestilential smell hung heavy in the air, an odour soon accounted for, for around the ravished camp lay at least a score of corpses, all in the same stage of decay as the one they had first passed on the plain. These appeared to be mostly old men and women, although here and there the smaller bodies of children could be seen amongst their slaughtered parents. Brown and the others drew a long breath as they gazed on this scene of murder.

" What a blessing it is," he said, " to know that all those wretches who did this are crushed into jelly underneath tons of rock."

" Yes," replied Morton in a low voice; " and for two pins I could find it in my heart to send that hoary old sinner there to keep them company."

This sentiment was a common one, and Columbus received some very savage glances, even Billy looking at him, and handling his carbine as though anxious to use it on the blackfellow. The old cannibal, however, was quite unconscious of the feeling he had aroused, and smiled sweetly as though he was showing them a highly interesting little exhibition.

"They must have killed and captured the whole tribe," said Morton at last. " No hope of finding Stuart now."

" I am afraid not. We had better look out a camp as far to windward of these poor wretches as possible," returned Brown.

Just then Billy whistled, and when his master

looked towards him, he indicated by a motion
of his head the direction of the hill with the
granite boulders on it.

A thin column of smoke was stealing up from
the back of it.

Morton whispered hastily to Charlie to take
Billy and ride round the foot of the hill to the
back of it, leaving the loose horses feeding about
on the green grass at the edge of the water.

He and Brown rode straight over the crest of
the hill, and underneath them saw the mockery
of a camp. A wretched remnant, who had
escaped massacre, they found huddled together
near some rocks. Two old men, about half a
dozen gins and children, and one young fellow,
badly wounded. Too startled and frightened to
attempt flight, they gathered timorously together.
Their fear seemed augmented when Charlie and
the black boy came up.

Brown dismounted and walked up to them. At
once a cry of surprise and pleasure broke from
the old men. They commenced jumping about,
shouting and laughing uproariously. Instantly
it flashed across the minds of the whites that they
were mistaken for a reincarnation or resurrection
of their countrymen who had formerly lived here.
Morton and Charlie left their horses and joined
Brown, when the conference was abruptly interrupted by the appearance of a riderless horse.
A difference of opinion had arisen between
Columbus and his steed when the others left

him alone, which resulted in the discomfiture of the native, who now followed limping after his horse.

The old men recognized their enemy at once. They stamped, raved, and spat at him; and one, picking up a spear, drove at him with such vigour, that only a nimble jump saved Columbus from being transfixed. For his part he returned their vituperation with interest, and the gins joining in, a perfect tempest of words ensued. Seeing that nothing could be done until they were alone, Morton told Charlie to go and round up the pack and spare horses, and, with Billy and Columbus, to take them some distance up the lake and camp on the bank, where he and Brown would join them.

Once their enemy was out of sight the blacks quietened down, and one of them commenced a voluble speech to Brown, addressing him as "Tuartee", from which it was evident that their first surmise was correct. After some trouble they made the natives understand that they were not going away, but were going to make a camp at the lake. They promised to return shortly, and rode away to join Charlie and Billy.

Everybody enjoyed a good swim in the lagoon, and ate a hearty meal afterwards. Brown and Morton then strolled back to interview the blacks.

CHAPTER XV.

The Fate of Columbus—Investigation of the Cave—
Stuart's Grave and Recovery of the Conclusion of
his Journal.

THEIR approach to the camp was greeted with
cries of "Tuartee", and they endeavoured
to make the old men comprehend that they
wanted to be shown the cave. Apparently,
however, the blacks, if they did understand,
thought that their guidance was quite unneces-
sary to a returned spirit, who ought to know all
about it. They imitated the gestures made by
the whites, and pointed with infinite politeness
the same way that they did, but that was all
that could be got out of them.

As they were both tired after their long and
dreary ride, they determined to start on an
independent search the next morning, and after
giving the blacks some trifling presents they
returned to camp. That evening they enjoyed
a meal of fine fish caught in the lake. It was
plain that some days must be spent in the neighbourhood, in order to thoroughly investigate
the caves, and find out if possible the fate of
Stuart.

"Well, we are here," said Brown, as he made
his bed down that night, "but I'm hanged if I

know exactly how we are going to get back again."

"No, we shall have to make a mighty long dry stage, for that last hole we were at will be dry by to-morrow."

"I think Stuart must have got across some other way."

They were soon all sound asleep; as no danger was to be apprehended from the poor wretches at the camp at the hill, no watch was kept. Towards morning Morton felt himself gently shaken by the shoulder; looking up he saw Charlie bending over him.

"I'm sure there's somebody prowling around the camp," he whispered. "I felt that funny feeling one feels, you know, of the presence of something not right about the place. I was woke up by a sound like a blow, or a stick breaking."

Morton sat up, looked around, and listened. All appeared peaceful and quite enough. The fires had burned down, and the light of the stars alone illumined the scene. The sheet of water alongside was unruffled, and reflected like a mirror the thickly-studded sky overhead. Not a sound could be heard, not even the cry of a night-bird or water-fowl. For some moments they both remained silent, listening, then Morton said in an undertone:

"Must have been fancy, Charlie. There can be nobody here but those poor wretches over the hill."

"No, it was not fancy," answered Charlie. "I am sure there was somebody moving about. You know I would not have roused you had I not been certain. Listen!"

Loud cries suddenly arose in chorus from the camp of the natives.

Brown started up.

"The devil!" he said, after listening. "That old Columbus at his cannibal tricks again. See if he is there, Charlie."

Billy and Columbus had made a separate fire, round which they were sleeping, coiled after the manner of blackfellows. Billy, aroused by the outcries which rung out clearly and distinctly in the still night air, now struggled to his feet, half asleep.

"Here's Columbus," said Charlie, giving the prostrate chieftain a good kick. "Wake up, old man!" he cried.

Columbus never stirred.

"There's something up," said Charlie, drawing back with a shudder.

Morton struck a match, as did Brown.

There was indeed something up. One glimpse was sufficient. Columbus lay dead, his skull shattered with a two-handed club which had been left beside his body. The shouts of the blacks were tokens of rejoicing at the return of his executioners with their work accomplished.

The whites gazed at the dead man in silence, and each felt slightly cold at the thought of the ease with which the whole camp might have been disposed of.

"Retribution!" said Morton at last. "He deserved his fate, but I can't help feeling sorry for the old villain. Billy, my boy, supposing that fellow had made a trifling mistake and tapped you on the skull in the dark."

Billy shook his head as though to make sure it was quite sound.

"No good this one country," he replied; "mine think it go back alonga station."

"Billy, your remarks, as usual, are to the point, and chock-full of sound sense," remarked Brown. "But we shall all feel better when the sun jumps up. Let's make the fire burn and have breakfast. It's not far off daybreak."

By the time the meal was finished the first rays of the sun were just visible. Charlie and Billy were sent after the horses, with instructions to remove the camp to one of the clumps of timber some short distance back from the lake, and then to take Columbus' body to where the victims of his tribe were lying—there they could moulder in company. Morton and Brown started in their search for the cave, taking the camp of the natives on their way. The killing of Columbus was only a just act of tribal vengeance. They did not intend, therefore, to let it interfere with their friendly intercourse with these natives, from whom so much valuable information might be obtained.

The blacks evinced no fear when they came to the camp, and greeted them in the most friendly manner. Not wasting any time in fruitless attempts at intercourse, the two men set out on their search. They were fortunate at the outset. They selected the two most imposing boulders, which seemed to answer best to the description in the journal, and on nearer approach a well-worn pad proved that they were on the right track. Squeezing through the narrow aperture described by Stuart, they found themselves in the cavern confronting the gigantic figure painted on the roof and side. Prepared as they were for the startling appearance of this form, they could not repress a certain feeling of awe as they gazed at it, and recognized at once that it was not the work of any Australian aborigines then existing on the continent.

A movement behind made Morton hastily turn round. One of the old men had silently followed them, and was standing a few paces away. Seeing Brown look curiously around after the first survey of the figure he advanced, and beckoning to him, led him to the side of the cavern where the light from a crevice above fell strong on a certain place. There, on the rock, had been carved with care this inscription—

"CHARLES NEIL STUART
CAME HERE, 1849.
DIED —— "

Then followed a date, scratched with a feeble hand, which they made out to be "1870".

With uncovered heads the two friends gazed sorrowfully and reverently at the resting-place of their unfortunate countryman. Although they had never really anticipated finding him alive, a feeling of sincere regret was uppermost at discovering their worst forebodings realized. They would have given much to have been in time to bring succour to the poor lonely man by the side of whose grave they knew they were standing.

The native again advanced, and putting his hand into a crevice in the rock drew forth a package, done up in the dried skins of some small marsupials, and furthermore protected by a casing of bark. This he gravely handed to Brown, who took it from him, but refrained from opening it at once. After a short scrutiny around, resulting in no further discoveries, they left the cave. Resting on the first convenient rock, they proceeded to inspect the precious parcel. The contents consisted of an old-fashioned double-barrelled pistol, a powder flask, a bullet-mould much dented and battered, and a roll of loose leaves of paper covered with faded writing. Together they pored over these leaves, which contained the conclusion of the castaway's life. They were in much better order than the contents of the pocket-book originally discovered, not having been subjected to such rough usage, and the narrative ran on without a break. The contents explained the presence of Murphy amongst the cannibals, the loss of the pocket-book, &c., and recorded Stuart's futile attempt to escape to the south, his meeting with the now exterminated tribe who lived at the foot of the mountain, and his return after repeated failures to penetrate the scrub and sand which cut him off from the settled districts. A gold discovery was also recorded.

They went back to their new camp, meaning to spend the rest of the day in copying out the journal, so as to insure its safety as much as possible. Morton dictated the narrative to Brown and Charlie, who made separate copies. Thus ran the story.

CHAPTER XVI.

The Continuation of Stuart's Journal—The Slaughter Chamber.

I AM now alone, and I know not whether my comrade is living or dead. It was a year after Kelly's death—by my reckoning, which I have kept by notches on a rock in the cave—that I went with three natives to a scrub about ten miles from here to get a peculiar kind of wood I was looking for to make bows of. For now that I had made up my mind I would never be rescued, I thought I would try to teach the natives the use of the bow and arrow, and we would lead them against this tribe whom they dreaded so and who killed Kelly, and perhaps obtain peace. There was no wood suitable near the camp, but from the description given by the blacks I thought I could obtain what I wanted in the scrub indicated by them. There was water there, and we stopped two days, cutting and dressing the saplings so as to make them lighter to carry in, for as we only had stone tomahawks it took a long time. On the evening of the second day we heard a gin wailing and crying in the distance, coming towards us. The blacks stopped their work and ran to meet her, crying out in the same tone. I knew something was wrong and followed them. It was sad news, awful news! The Warlattas, as the hostile tribe was called, had attacked the camp at night, had killed and wounded many, and carried off a number of prisoners—amongst them Murphy, who was a heavy sleeper and had no chance to defend himself. I knew that these Warlattas were cannibals, and that the prisoners they took away were probably eaten.

"We got back to camp in the middle of the night, and the next morning I tried to get the men who were left to follow me after the cannibals, but they were all so cowed they wouldn't, although I showed them the pistol and fired it off. I tried to track the enemy by myself, and

if I could I would have followed them, but I lost
the tracks and nearly died of thirst. The Warlattas had taken nearly all the few things we
had saved, including my pocket-book; these few
sheets I am writing on were picked up about
the camp.

"1853.—That is my reckoning. All this time
I have written nothing, as I wanted to husband
my paper, and I had little heart after Murphy
was taken away. I made the blacks build a
place with stones—a sort of barricade to sleep in
at night,—and it was lucky I did, for the Warlattas came again; but, thanks to the barricade
and my pistol, we beat them off without losing
a man, and now the natives have great confidence,
and I think will beat them again.

"I often tried to get them to follow me to where
these people lived, as I thought Murphy might
be alive and I could rescue him, but they seemed
to be horribly frightened at the thought and
refused always. On examining the bodies of
those that had fallen, I found them all marked
the same way with some sort of pigment, a red
smear on the forehead and a white triangle on
the breast. This, and something in their appearance, led me to consider if there was not some
connection between the figures in the cave and
this strange people. Thinking long over this, I
explored the cave thoroughly, both it and any in
the neighbourhood, and finally it led me to the
strange discovery that has caused me to write my

journal once more, in the faint hope that some day it will be found and read by civilized man.

"Searching around the cave containing the painted figure, I found an aperture which apparently ran for some distance. It was on the ground, the rock coming to within about two feet of the sandy floor, and on stooping down it seemed to me that I could feel a current of fresh air passing through. On inquiry I found that none of the blacks had been into the opening, as they had a superstitious dislike—scarcely, however, amounting to dread—of the cave. The aperture was too low to easily admit me, so I got a slim young fellow to explore it. He soon crawled back, saying there was another big cave beyond, but too dark to see anything. I got some more boys up and set them to scoop the sand away until the opening was big enough for me to pass in. We took fire and bark and wood with us, and when we emerged in the gloomy cavern beyond we immediately kindled a fire. As the blaze arose and illuminated the recesses of the cave a shriek of terror burst from my juvenile companions, a wild cry of " Warlatta! Warlatta!" and in an instant they disappeared like a bevy of black rats underneath the rock where we had entered. I looked around in surprise, but soon divined the cause; on the opposite side appeared, drawn in white on the wall, a large triangle, the sign ever associated in their minds with murder and rapine.

"Heaping more wood on the fire, I advanced and examined the surroundings. Underneath the triangle was a huge block of yellowish-white sandstone, but its purity was marred by a horrible reddish stain which marked one of its sloping sides. Its purpose flashed on me at once —in some old time it had been used as a sacrificial stone. The fire now blazed up merrily, and I had ample light for my researches. The smoke disappeared through crevices in the roof, and the ventilation seemed excellent. Marks of old fires were visible all over the floor, which was of white sandstone with the same reddish stains visible in places. Searching more minutely I found in one corner a knife or dagger, made of steel (since then I have found it to be tempered so skilfully that the edge can scarcely be turned by the hardest rock). The handle, if it ever had one, had disappeared through age. In addition, there was a broken ring of the same metal, seemingly part of a chain, and on the walls were characters in red, but of no written language that I could remember. This was all that I saw on my first visit.

"Voices at the opening told me that the natives had recovered from their fright, and were in search of me. I called to them, and emboldened by my voice and the firelight some of them crept in and joined me. I found out that they had no knowledge of this chamber, and in hopes of finding another I set them hunting round for any more openings that might exist, but none could be discovered. Whilst so engaged one of them brushed against the stone altar, and immediately it commenced rocking, whilst a squeaking, piercing scream, like a human being in intense agony, thrilled us all with horror. The blacks threw themselves on the ground, and it was a few moments before I could summon up courage to approach the stone and examine it. The rocking was gradually ceasing, and the shrieks grew fainter as the motion ceased. The stone I found to be most beautifully poised, so that the slightest touch started the oscillation. As to the machinery that produced the screaming noise, that I could not investigate without capsizing the stone, which evidently weighed some tons. For a moment I shut my eyes, and seemed to see once more the hideous drama that must have been many times enacted in this chamber of death—the savage priests, the manacled victim, the streaming blood, the trembling captives, and the harsh shrieking of the rocking stone adding its awful voice to the groans of the dying man and fading away into silence with his last cries. What horrible ingenuity had devised such added terrors to the scene? By degrees I got the blacks out of their fright, but it was amusing to see the celerity with which they disappeared as soon as I gave the signal.

"1862.—I have made a great effort to escape, but am forced to come back here to die. The blacks had told me on two occasions that rather to the west of south there was water within reach of a long day's journey; but as this was only leading me further into this uninhabited wilderness, I had never had the curiosity to go there. It now struck me that from there I could possibly get round the end of the great sandy desert, and perhaps find an easy road back to the settlement, which must have pushed out towards me since I have been buried here. I had succeeded in teaching the blacks the use of the bow and arrow, and to build tolerably safe huts to sleep in. The Warlattas had attacked us twice since the first time we defeated them, and on both occasions had suffered great loss, whilst we had not a man wounded. For years now they have not dared to come, and I think the bows and arrows have frightened them. Moreover, my natives have no longer the terror of them they had formerly, and feel confident in repulsing them. Under these circumstances I felt that I could venture to leave them, for I did not like the idea of their becoming once more a prey to this horrible Warlatta tribe. One of the old men who had been to the water before, and a fine young fellow named Onkimyong, accompanied me. I fully explained to the blacks what to do if the Warlattas turned up again, and promised them soon to return; for if I succeeded in getting away, I meant to come back with a party to thoroughly examine the caves and root out the Warlattas for good. Strange, the blacks have no repugnance to going anywhere west or due south, but to the eastward they will not go.

Our journey during the first day was over treeless country well grassed, although at times we came across patches of the prickly grass, proving that we were not very far from the edge of the sandy desert. We did not reach the water that night, but as we had brought a couple of coolamen[1] full, we did not trouble to press on. Next morning we arrived there early in the morning and found it a long narrow lagoon, the water being of a milky colour. Around this lagoon were many camping-places of the natives. I asked the old man if he knew this tribe, and I found that he had met some members of it once; they were friendly, not like the Warlattas."

CHAPTER XVII.

Continuation of Stuart's Journal—A Hopeless Situation.

WE stayed at the lagoon all day, and in the evening, fortunately, a party of the natives came in. They were timid at first, but the old man and Onkimyong could make themselves understood, and they gradually gained confidence. They had never seen a white man before, al-

---

[1] Vessels chopped out of the soft wood of the coral-tree by the natives; and used for carrying water in the dry country.

though I am now pretty well burnt black by the sun. My two natives showed off their bow-and-arrow shooting with great pride. They told the others how the Warlattas, who seemed to have turned their attention to the new-comers also, had been beaten off and killed.

"These natives explained that they lived on a creek to the south-east, and when I heard that I made sure that I should at last escape. When the old man found out where they came from and that I intended to accompany them, he would not go any further, and nothing could induce him to alter his intention of going back. Onkimyong, however, who was very fond of me, and being young had not so much superstition, said that he would stay with me and go wherever I went.

"The blacks were on a hunting expedition, and had come to the lagoon on purpose to fish; so we remained there a few days and the old man returned to the lake.

"When we started we went to the south-east, and the country rapidly changed its character, becoming scrubby and barren. That night we camped at a salt lake, obtaining some water, slightly brackish for drinking, from a native well dug some distance back. Next day our course was through wretchedly poor and barren country. When we rested for a time I noticed an outcrop of quartz; my position in the party had ostensibly been that of geologist, and I went over to examine it, for before we left there had been some vague rumours that gold had been discovered in the southern part of the colony. I broke up some of the stone with a large one, and found that it was auriferous. This discovery did not elate me in any way. If I had found a mountain of gold, of what value would it be to me?

"Continuing our journey we reached water again that night, apparently a small soakage spring. The blacks told Onkimyong that we should camp at a small creek the following night with some brackish water in it, and that the next night there was water in a clay pan, and the following night we should reach their main camp. This proved to be the case, and we found their home to be on the bank of a fine creek, running round the foot of a tall hill. I now looked upon my escape as secure, for surely this large creek, well defined and supplied with water, must run down south to settled country, and I could follow it easily. Alas! I was doomed to disappointment!

"The Warlattas had not been seen for some time, and, unluckily for them, they selected the second night after our arrival for an attack.

"It was brilliant moonlight, and the blacks were holding a corroboree in our honour, when one of the gins shrieked out that the Warlattas were on them. The fire-sticks were visible coming on swiftly, and they had evidently reckoned on taking the camp by surprise. I had been very careful of my ammunition, but I thought I could spare one charge. I called to Onkimyong, and told him to tell the blacks not to be frightened; then, as the Warlattas approached, shouting and yelling, I fired straight at them. The effect was instantaneous—the onslaught stopped at once. It must have completely surprised them to find themselves suddenly confronted by me in this new place. Before they could recover from their surprise Onkimyong and I were at work with our bows and arrows. This completed the rout, and they turned and ran; Onkimyong shouted to the natives and rushed in pursuit, followed by some who had recovered from their terror. I did not go with them, but I think they did good execution.

"There was great rejoicing over this defeat of their enemies, and I felt very glad that the attack should have been made when it was. Seemingly, since we had beaten them off at the lake they had devoted all their attention to this poor tribe. The next day I ascended the mountain, and from the top saw that in the direction I wanted to go there was nothing but a vast scrub. The creek, too, seemed to disappear soon after passing the mountain; and this I soon found out was the case. It ran completely out in a sandy waste of scrub. The blacks asserted that it never re-formed, and that there was no water either to the south or east, and that nothing lived there but snakes. I tried over and over again but always had to return, half dead with thirst and fatigue. One old man said that he had heard of a big rock down south where there was a hole with water in it. But this I imagine was only a tradition, as from the top of the hill I could discover no sign of it, and wherever I penetrated I found always the same arid and barren scrub and sand. Being thus disappointed in my efforts south and east, I thought that I might follow the creek up and come to some available strip of country. Judging by its direction the creek, if it headed far enough away, must be east of the prickly grass desert.

"As the Warlattas always came down the creek I could not induce one of the natives to accompany me. Even Onkimyong was afraid to face it. With little care about my fate I therefore started alone. I followed the creek for a long distance, finding it well watered, and that a beaten track ran beside it. This turned off, and on following the creek further I found that it ran out; I therefore returned and followed the track. In course of time this led me to a swamp of great tea-trees which it skirted. After following this swamp half-way round the track left it and went amongst some rocks. They were basaltic, and in a short time they closed in in a perfect wall and I lost all trace of the track I had been following. Again and again I tried to find it, but the rough basalt cut my feet to pieces and the track could not be followed over the rocks. I had to rest for a time to get my feet well, and fortunately there was plenty of game about the creek, which apparently re-formed on the northern side of the swamp. I now determined to follow this creek up again, and did so, until at last it died out in a desert forest. At one place I saw a number of trees marked, apparently by the Warlattas. I made several excursions east of the creek, but I was always confronted by a dense and impenetrable scrub."

"Poor fellow!" said Brown at this point. "Fancy his being so near his companion Murphy and yet to miss him."

"I can well understand his inability to get along through those basalt rocks, but I don't understand how he did not see the Warlattas' track at the lagoon of the marked trees."

"If you remember," replied Brown, "the track was not very plain close to the lagoon."

"I had at last to give up in despair" (went on the journal), "and make my way as best I could back to the mountain. How long I was away I cannot say, for I lost count. It seemed to me weeks, but I think it was about a fortnight.

"I was now thoroughly convinced of the hopelessness of my situation, and determined to return to the lake and finish my weary life amongst the tribe there, devoting my time to teaching them what I could.

"Onkimyong was delighted to see me back. I rested for some time, as I had two or three things to do before leaving. One was to show the natives how to build a stone barricade, and the other was to inscribe my initials in some place where it was bound to be seen by any whites who might hereafter come. I selected a place at the foot of the hill for the barricade, and set the blacks to work, under the superintendence of Onkimyong. From its position and altitude I concluded that any whites coming to the place would naturally ascend the hill to obtain a good survey of the surrounding country; I therefore inscribed my initials and the date of the year on a rock on the summit, doing the work with the aid of the knife I had found in the cave. I lingered on for some time longer in the hope that the Warlattas would make another attempt, and this they did the night before we were going to leave.

" Fortunately their approach was discovered in ample time, and I had my men all concealed behind the barricade. The Warlattas approached very cautiously, not with the confidence of their first attempt. We allowed them to come pretty close, and then commenced to play on them with our arrows. As soon as I saw them waver and halt, I gave a signal agreed upon, and the natives swarmed out and attacked them with their clubs and spears.

" There seemed to be no hesitation this time,

with one accord the Warlattas fled. The pursuers did much more execution than the first time, as they had a better start. I hope now the Warlattas have received another check.

"Onkimyong and I started back the next day, we followed our tracks back again, as I felt curious about the gold-bearing reef. When we came to it I examined it thoroughly, and I then found that it was, what seemed to me, of fabulous richness. I laughed aloud. Here was I with a fortune at my feet, and it was of no more value to me than worthless flints. It was the very mockery of riches!

"In time we arrived at the lake, and met with a great welcome, as they had given us up as lost. I had been in fear that the Warlattas, finding I was away, might have attempted another assault, but they had not put in an appearance.

"I have now quite relinquished any hope I had left of finding my way back by my own exertions, and can only pray that some other exploring party with better fortune may come here before I die."

CHAPTER XVIII.

Conclusion of Stuart's Journal—Examination of the
Slaughter Chamber—The Ancient Australians.

I HAVE made no other discoveries since my
return, and all the efforts I have spent in trying to decipher the inscriptions have been in vain.
I can only conjecture that these relics are of
great antiquity, and that the belief and some of
the rites, notably cannibalism, survive amongst
the Warlattas, who are mixed and degenerate
descendants of the ancient race. I have very
little paper left, and that scrap I must keep for
any necessity that arises. If anybody finds this
let him take a copy of the inscriptions, for there
may be some men in the world who can decipher
their meaning.

" 1865.—I have devoted myself to bettering
the condition of this tribe, whom I may say I
have adopted. I have taught them to build
better huts, and clothe themselves partly in
skins. The Warlattas' inroads have been abso-
lutely stopped. They have learnt to cultivate
yams here, and some of the young men understand written signs. One thing I could not induce them to do with all my influence, that is, for
a party of them to go east with me and find out
the track by which the Warlattas cross the

sandy desert. Some superstitious feeling I cannot overcome will not allow them to do this.

"I might have done much more, but latterly I have been nearly crippled with rheumatism. I have instructed the natives to bury me in the cave under an inscription I have cut—my name and the date of my arrival. When I feel myself near my end, unless I die by accident, I will try and inscribe the last date on the stone, it will stand for my death year. If my companions had lived, we might have worked our way back to the settlements, but alone—it was hopeless to attempt it. I know my end must be near, nor am I sorry, for I have outlived all hopes of succour. I thank God that though I have lived so long amongst these savages, I have not sunk down to be one of them in their habits, but rather have taught them better things. To the white man that finds this I leave the greeting and the blessing I would have given him in life."

. . . .

"What would I not have given to have got here in time to rescue him!" said Brown. "He was a man worth saving."

Next morning they took some more presents to the natives at the hill, and the old men went round with them and showed them the roofless stone huts, the dismantled barricade, and the remains of other improvements now all in ruins. The death of Stuart seemed to have been a signal for a return to their old habits of life, his

M 64

MORTON AND HIS PARTY EXAMINE THE SLAUGHTER CHAMBER.

stay amongst them not having been long enough to make a lasting impression.

Even the bows and arrows had disappeared; and it was evident that the Warlattas had resumed warlike operations with a success resulting in the almost complete extermination of the tribe. Morton endeavoured to explain to them that their enemies were dead, but it was doubtful whether the old men comprehended him. An immediate incursion into the inner cave was determined on, and, provided with candles, the party soon found themselves at the opening. The sand had worked in and somewhat blocked up the space, but this was soon sufficiently removed to enable them to wriggle underneath like snakes. Half a dozen candles served to brilliantly light up the inner chamber, and there, with startling distinctness, shone out the white triangle over the sacrificial stone. Brown started the stone rocking, and immediately the shrill, half-human screams echoed through the cave, much to Billy's discomfiture. No examination could detect the trick that caused the sound, nor could the presence of the stone be accounted for except as a most singular freak of nature.

"I have it," said Morton at last; "the stone was part of the rock and has been cut away underneath. It must have been an awful job, but that is how it was done."

"And about the squeaking machinery?"

"That's more than we can find out without shifting the stone, and that's a job I am not on for unless we stop here a month or two and chip it in pieces."

"A charge of dynamite would shift it; and next time I go exploring I'll carry some," replied Brown.

"What do you think the thing was made for?" inquired Charlie.

"Well, I've just had an inspiration," said Brown. "You know amongst some nations it is a matter of religious belief almost that you must make your enemy howl when you have got him down. Now, perhaps some of the poor devils who were cut up on this stone declined to sing out, or fainted, or for some reason did not furnish amusement enough—the stone was set rocking to fill out the programme. What do you think?"

"I think it most likely, and is an instance of devilish cruelty on a level with their other proceedings."

"I suppose poor Stuart searched this place so thoroughly that we need not expect to find anything fresh," said Charlie. "But we may as well have a look. Billy, you have got sharp eyes, just use them."

While these two were investigating the walls and floor, Morton and Brown took a careful copy of the hieroglyphics and a sketch of the cave, showing the position of the sacrificial stone and triangle. By the time they had finished they were ready for their mid-day meal, and returned to camp for it.

"There's no doubt," remarked Morton, "that what we have just seen are relics of an ancient people, but what I can't understand is, why, if they were civilized enough to wear dresses, and to have a developed religious belief—savage as it was—" ("No worse than the Carthagenians," interjected Brown), "to know how to obtain iron and temper it, that they did not build permanent buildings, the ruins of which would remain?"

"Mud, my dear fellow, mud," replied Brown. "Remember the nations who have disappeared off the face of America, and can only be traced by their pottery and burial mounds. Why, the gorgeous cities of ancient Mexico were built of mud bricks, which go back to their mother earth, once the domiciles they form are abandoned."

"But their smelting-works for manufacturing iron?"

"There you have me. But we must try and find that knife; perhaps they buried it with Stuart."

"Billy got something from one of the old men, but I don't know what it was," said Charlie.

"Billy! What old man bin give it?" asked Morton.

Billy grinned, and produced from the inside of his shirt the knife mentioned in the journal. It was a curious-looking blade about a foot long,

broad and somewhat curved. Even after the
work Stuart had done with it in carving on the
rocks, it was as sharp as an ordinary knife.

"That's a Malay weapon," said Brown, after
examination. "Whoever our ancient Australians
were they came from the north. I suppose we
must wait until we get the writing deciphered,
if there is a man clever enough to do it."

Looking closer, they found that the blade had
the mysterious triangle engraved on it. This constantly recurring symbol led to much speculation
as to whether they were not an offshoot of Freemasons, who in some remote time had wandered
into central Australia; but as Charlie ingenuously
reminded them that these fellows had not built
anything, the theory had to be discarded.

"What's to be done next?" said Morton. "I'm
for going to find Hentig's grave, if possible, and
recovering the papers buried there."

"Yes, and then go back round by the way
Stuart went to the mountain tribe and track up
the gold reef."

"Not a bad idea, anyhow. I think it will be
the safest way to go home."

That afternoon, Billy, with the aid of one of the
old men, found a canoe not far below the surface,
and brought it up and bailed it out. Then it
transpired that on the last onslaught of the
Warlattas they had sunk all the canoes. The
one recovered was a large one, fitted up with
outriggers, and leaked but very little. Charlie

soon improvised a mast by lashing two spears together, and with a blanket for a sail announced himself ready to face the dangers of the deep. Morton agreed to join him in his voyage of discovery round the lake the next morning, but Brown preferred to stay and continue the investigation of the cave drawings.

Next morning there was a gentle breeze blowing, and Morton and Charlie were soon afloat and off. Brown wandered over the hill, telling Billy to try and make the blacks understand the catastrophe of the burning mountain.

Several lesser caves attracted his attention, but only one seemed to promise any result. To this one he devoted himself, and after some trouble found some inscriptions resembling the former one in character, but differing in the arrangement of the letters. In this case they were placed perpendicularly in two parallel lines.

After copying the inscription, Brown stood in thought for some time, mechanically thrusting a yam stick he held in his hand into the sandy floor of the cave. The soil over the bed rock in this cave was apparently only a few inches in depth, but suddenly he was roused from his reverie by the yam stick going down more than a foot without meeting with any opposition. Sounding hastily, he soon found that a trench extended at right angles to the rock, immediately under the inscription. Going outside he shouted loudly, and Billy and some of the blacks came running up. He set them to work to clear all the sand away from the trench and its neighbourhood with their hands, and it was soon visible artificially cut in the rock. It was about three feet long and a foot broad, so the work of clearing it out did not promise to take long.

With so many hands going, the depth of three feet was soon reached without anything being discovered; then the fingers of the workers came in contact with something hard, and very soon a sheet of metal was disclosed, cut exactly to fit the hole. Brown at once recognized it as resembling the gongs used at the burning mountain. Greatly excited, in spite of his usual assumption of calmness, Brown inserted the point of the yam stick under the metal and prised it up.

CHAPTER XIX.

The Grave of one of the Unknown Race—Morton's Departure—Charlie falls Sick.

THE shadow cast by the sides of the hole was too dark for Brown to see at once what sort of treasure he had unearthed, but a nearer inspection did not afford him any more satisfaction. Apparently there was nothing beneath the sheet of metal but a deposit of damp, mouldy earth, emitting a pungent smell of a most repulsive nature. Brown drew back somewhat sickened, and told the natives to clean out the trench while he went out for a breath of fresh air.

In a few minutes he returned. The heap of mould alongside the now empty hole told him that the work was finished. The blacks were on their knees, eagerly examining some objects they had found amongst the contents of the trench. Billy handed them to Brown. The first was a chain of small steel links and most beautiful workmanship, bearing as a pendant a tiny triangle formed of the same metal. A metal plate nearly a foot square, covered with hieroglyphics similar to those inscribed on the walls of the cave, was the next thing he examined, and then came a dagger resembling the one already discovered. This, however, had a handle, or what appeared to be one, made of finely-twisted gold threads wound tightly round the haft. This was all. Brown puzzled over these relics for some time, and then strolled to the crest of the ridge to see how the voyagers were getting on. Apparently they had experienced what is known as a soldier's wind, for the canoe was coming back with a flowing sail. Brown and Billy walked down to the shore of the lake to meet them. The trip had been uninteresting; the lake was exceedingly shallow everywhere with the exception of the end where they were. There it was deep and permanent. Brown told them of the discovery he had made, and they revisited the cave together.

" It's my opinion," said Morton after lengthy examination, " that it is a grave, and this mould all that there is left of a body—probably burnt before being put in the grave. The necklet, plate, and dagger are the ornaments and weapon worn by the man at the time of his death. He must have been some great priest."

" The same conclusion that I have been working round to," replied Brown. " That plate is a priestly breastplate, like those we read of in the Old Testament."

The question of visiting Hentig's grave was one that was now discussed. It was settled at last that Morton and Billy would make the trip, leaving Brown and Charlie to continue the investigation of the caves, and also to come out with relief in case Morton was overdue. By careful computation they thought they could pretty well guess the course pursued by the three white men from the last water to the lake. With two horses packed with water they reckoned they could get out and back without much trouble, even if there was no water at the marked tree.

Morton and Billy therefore started early the next morning, for although the night would have been cooler to cross the spinifex desert, they might have missed some important indications in the dark. Four days was the utmost limit Morton allowed himself; Brown was then to start on his tracks, as something would have probably happened to the horses.

Left at the lake, Brown and Charlie devoted themselves to a searching examination of the locality. Several trees marked with an anchor, similar to the one at the lagoon camp, were discovered, evidently the work of Murphy, who had seemingly appropriated the symbol. A mark resembling a rude **K** was seen once or twice, and they took it to stand for Kelly's handiwork.

They sounded the caves all over but without any more success, and at last concluded that they had found all there was to be found.

The sun was high the morning of the third day when Brown returned from his swim in the lagoon, and, to his surprise, found Charlie still sleeping, with a hot flush on his face. Brown aroused him, and the boy sat up and looked vacantly around, recovered himself after a bit, and proceeded to get up. He refused to eat anything, but drank a good deal of tea. Brown watched him anxiously.

" What's the matter, old boy?" he asked.

"I don't feel up to the mark. I had a wretched nightmare last night; it kept me awake afterwards until nearly daylight, so that I overslept myself."

" I feel off colour too," replied Brown. " Last night I could have quarrelled with my own shadow. I hope we didn't release any evil spirit from that grave."

"Don't say that," replied Charlie, "for that is just what I dreamt. Strange that you should think the same."

"Tell me about it, sonney," said Brown. "We'll soon fix up any intrusive old ghost."

Charlie, as he could see, was upset by something, and Brown felt uneasy, as the thought of sickness overtaking one of their party became patent to him.

"I dreamt," commenced Charlie, "that I was on the edge of this lake. I was alone, and very frightened with a quite unnatural terror. I thought you had both gone away and left me. I tried to cry out but could not, and turned round, for I felt some fearful thing was approaching me from behind. True enough, the great figure from the cave was there, looking at me with terrible eyes. On its breast was the plate we found in the grave, and—I could read the characters written on it."

Charlie paused.

"What was the writing?" asked Brown, interested by the boy's earnestness.

"I can shut my eyes and see it now. It ran— '*The Spirit of Evil is everywhere. Worship then the Spirit of Evil only, and do his behests.*' As I looked and read the figure smiled mockingly at me, and I woke up in a cold perspiration, and could not sleep again."

"Charlie, my boy," said Brown, "we will discuss your dream by and by. Meantime, I am going to mix you a dose of quinine and brandy; you've got a touch of malarial fever coming on. Now I'll fix up a bough-shade for you; it will be cooler than the tent, and you must keep quiet all day."

Brown soon had a good shade of boughs erected, and making up as comfortable a bed as he could for the sick lad, he stopped with him all day. Charlie was very feverish, but towards sundown he fell into an uneasy sleep, and Brown went for a stroll up and down and smoked his pipe. "This is a lively look-out," he mused. "I hope those devilish old rites don't mean to claim another victim. Dreamt we both went away and left him!" Brown's eyes grew moist as he thought of the possibility of the words coming true in one sense, and Charlie being left in a solitary grave by the side of the lake. "If Morton does not turn up to time what a fix I shall be in, for I can't go and look for him."

Charlie passed a restless night, and towards the middle of the ensuing day he became delirious. This was the margin of Morton's return, but the sun set, and Brown strained his eyes in vain across the plain.

Charlie's delirium was at its height that night. Always he raved of the great figure in the cave standing over and threatening him.

The fifth day passed and still no sign of

Morton, and Brown was nearly distracted at the thought that his friend was in some difficulty, expecting him to come to his relief, and he could not leave the sick boy for an hour. He passed the night once more by Charlie's side, trying to soothe him and listening to his incoherent mutterings. It was about three o'clock in the morning when Charlie, who had been quiet for some time, dropped off to sleep. The silence that ensued was broken by a sound more grateful to Brown than anything else he could possibly have heard —the distant sound of a horse-bell. There was no doubt about it, the horses were coming rapidly across the plain, and evidently being driven, the bell having been loosened as is generally done when travelling with pack-horses at night. On they came at a sharp trot, for they were no doubt thirsty and knew where they were coming to. Brown listened till they ran down the opposite bank and commenced drinking, then he knew by the voices that Morton and Billy were both there, and called across to them.

"That you, Brown?" came back in reply.

"Yes; I could not go out as I promised."

"Glad you didn't, as it happened. But what's up?"

"I'll tell you as soon as you come round, but come quietly."

Brown walked a little distance to meet them, and they unpacked the horses where they met so as not to disturb Charlie, who was still sleeping.

Brown told Morton in a low voice, and they went into the tent together. Charlie was muttering in his sleep, and it was still of the figure on the wall. They went out and sat by the fire, where they could hear the slightest sound, and Morton told Brown all that had passed since he left.

CHAPTER XX.

Morton's Trip to Hentig's Grave—A further Discovery.

WHEN we left the lake that morning," began Morton, " we passed, of course, over a few miles of good downs country before we came to the edge of the desert, then we had nothing to do but keep straight on the course we had selected until night, when we had to camp in the spinifex. Next evening we came straight to the water-hole. It was a splendid fluke, and I never expected such luck. There was a good supply of water in the hole and fair grass, so we were not badly off. I found the tree where Hentig was buried, although the cross that Stuart cut had nearly grown out. The powder-flask, however, had been only buried a short distance in the ground, and as the bush rats had dug it up it had rusted and rotted almost to pieces. We found some of the contents, but the writing is almost if not quite illegible, as the paper has been soaked into pulp two or three times. As far as I can make out, though, it was written by Leichhardt himself, and as such will be valuable. Now I come to what detained me. Billy, whom I told to look on all the trees about the water-hole for marks, found a couple of initials on three of the trees. Of course the sides of the bark had grown together, and could only be traced like a crack in the bark. I made these initials out to be L. F. If you remember, Stuart mentions that the two men who were lost never reached this water, but one of them must have got in to it after Stuart left. I say one of them, because the same initials were repeated. I lay awake for a long time reasoning the thing out. When Stuart and the others left there was no water in the hole, they used up the last of it; therefore, this man must have come back after rain had fallen, probably some months after, for he could not have stayed about cutting his initials on trees without water. He must certainly have found other water. Where was it? If I could find it I might get further information.

"In the morning I sent Billy back along our tracks with a note in a cleft stick, which he was to stick up right on our tracks. In the note I told you to come on, as there was plenty of water in the hole. Billy had instructions to ride till noonday, then stick the stick in with a handker-chief tied to it, and come back again. All this he did, meanwhile I went fossicking about. The hole was on the edge of the plain, the same as the lagoons we were camped on so long. I went north first, and presently was able to find a kind of water-course skirting the forest. Once or twice I came to holes that should hold water for some time, but they were all dry. At last I was rewarded. I came to a fair-sized lagoon, with ducks and other waterfowl on its surface. There was no sign of the place ever having been much frequented by natives, at least I only saw some very old camps at first. At the end of the hole opposite to me was an old shell of a gum-tree, one of these desert gums that grow to a very old age and become quite hollow. As I looked at this I saw something move, and then, looking more intently, I could see some blue smoke stealing up. Naturally I made for the place as soon as possible, wondering if I was going to find a living white man. When I reached the spot I found a small fire burning, and in the tree an ancient old gin was squatting. She was almost blind, and could just make out that something was moving about, for she snarled and struck out feebly but viciously with a yam stick. Oh, she was a cheerful old lady!

"I was puzzling my brains how she lived, for she seemed too feeble to move and was too blind to see anything distinctly; for all that there were plenty of feathers around the camp, although she could never have caught the birds.

Presently the old dame, still growling to herself, crawled out on her hands and knees to the fire, to which she seemingly guided herself by the sense of smell. Here she squatted down, and I hung my horse up and went to look at her lair in the tree. There was an iron tomahawk shipped on to a black's handle, a single-barrelled percussion shot-gun, and all sorts of little odds and ends. Evidently this survivor must have found a lot of things abandoned at the old camp at Hentig's grave. Just then my horse shook himself, making the saddle rattle and jingle. On hearing this the old gin set up the most awful screaming you ever heard. When she quietened down someone called out to her from a short distance away, and looking in the direction, I saw a light-coloured nigger limping towards us. He stopped and had a good look at me before approaching; I held up both my hands as a sign of amity, and he came on pretty fearlessly. I then saw that he was a half-caste and a cripple. One leg had been broken in childhood and he had grown up with it shorter than the other, and much distorted. That he was a descendent of the missing white man I had no doubt, but the old gin appeared too old to be his mother. He came up to me, and when I spoke to him seemed greatly pleased, and pointing to himself said "Lee-lee" two or three times, indicating that that was his name. I gave him my name in return, which he soon picked up. He then led me to the tree, and taking the gun out put it to his shoulder and cried out in imitation of a report, showing that he had often seen it used. I pointed to the feathers, and he showed me two or three light boomerangs and a fishing-net. I tried to find out if he knew anything of the lake, but he seemed quite ignorant of its existence. I imitated death to find out what had become of his father, and he led me to a place where he indicated he was buried, but I could see no sign of a grave. He knew two or three words of English,—water, gun, tree, and bird; and I think he must at one time have known much more, but had probably lost them since his father's death, who, I think, must have been a man of the same intellectual calibre as Murphy, and quite uneducated. He explained by signs that his leg had been broken by a branch breaking when he was climbing a tree as a little fellow. Suddenly it struck me that there might be more in the family, and after some trouble he led me to believe that there were, but they had gone away with the tribe to the north. He limped heavily to show me that he had been left behind because he could not travel fast enough, and I concluded that the old gin who had been left was no relation of his, but had stayed behind from infirmity. Lee-lee seemed very active and clever, considering the disadvantage he laboured under, and I made up my mind to bring him in to the lake. He, however, did

not seem anxious to go at first, but I showed him
the empty gun, of which he was very proud, and
made him understand that if he came with me I
would make it alive again, which he seemed to
approve of highly. Fortunately there were some
boxes of caps amongst his belongings, as ours are
all breech-loaders. He knew the hole where
I was camped, and intimated that he would come
down in the morning. All this took some time,
and it was late at night when I got back to camp,
where I found Billy blubbering under the impression that I was not coming back again.
Lee-lee duly turned up the next morning, and
Billy tried to talk to him, but did not make
much of a fist at it. He said that he thought
some of the words were the same as those used
by the lake natives. Lee-lee knew nothing of
the Warlattas; and, if you remember, the defunct
Columbus told us that there were no natives to
the north. I gave him sugar, which he highly
appreciated, then we saddled up and went up to
his lagoon, which I reckoned to be nearly fifteen
miles away, so that he must have started pretty
early in the night to come to us. I fixed him up
on one of the pack saddles, and he got on very
well considering.

"The forest bears to the westward, and I intended to start straight to the lake from Lee-
lee's lagoon. The difficulty was about the old
gin. If left behind she would starve, and I did
not see how we were to get her across the desert.

I explained the dilemma to Lee-lee, who seemed to understand. Suddenly a bright idea struck him. He picked up his nulla-nulla and indicated that the easiest way to settle the question was by knocking her on the head. He appeared rather surprised at my objecting somewhat angrily to this simple and easy method, and I am not sure that Billy did not agree with him.

"I thought that as your ideas have been so brilliant lately, that we might devise some means of getting the old gin safely across the desert, and making these fellows and Lee-lee friends, so that, if we make up our minds to take him back with us, the old woman would not starve, for there is plenty of food about here. I gave Lee-lee to understand that I would be back in three days; but, of course, that is knocked on the head, we must get Charlie well first. Now, old man, you've had no sleep for two nights. I will sit by Charlie, and you can have a snooze until daylight. Watching is far more tiring than riding."

As Brown really felt somewhat tired out, he adopted the suggestion and retired to his blankets.

Charlie was no better in the morning, and Morton felt quite cast down at the sad fate now looming, only too plainly, before his young relative, for whom he entertained a great liking. About mid-day Brown suddenly arose as though filled with a new idea. He went off in the direction of the hill where the blacks were camped, and Morton did not see him for nearly two hours, then he said he had only been over to see how the niggers were getting on, and was silent and abstracted until darkness fell, when he persuaded Morton to go and rest while he kept watch by the invalid.

Morton, who had been riding and watching all the night before, slept late. When he awoke he saw Brown standing by the fire smoking.

"How is he?" he asked, as he took his towel to go for a swim in the lake.

"Better, I think; he has been sleeping quietly all night without talking. When he wakes up he will be sensible, I think."

"That's good news," returned Morton. "I shall be glad to see the boy up again. What a blessing it was that this thing happened in a good camp with plenty of game of all sorts! We must feed Charlie up well now."

Brown puffed on, looking steadily in the fire.

"I suppose you will think me no end of a fool for what I have done," he went on at last. "But I have not been able to help associating Charlie's illness with my opening that grave and taking out that devilish old plate. I have had that same dream that Charlie had, and could plainly see the plate and the inscription on it about the Spirit of Evil. I believe if I had not done what I have done not one of us would have got back alive."

" What was that?" asked Morton.

"Took it back to the grave yesterday and filled the whole thing up, and now Charlie is going to get better. What's the verdict?"

" Well, I was going to call you a thundering old idiot, but in view of the circumstances I won't. It must have been a tribe of devil worshippers who originally squatted down here."

" That's a weight off my mind. I thought you would have cut up rusty, for there's no doubt of the value of that relic. But we have copies of all the inscriptions."

Charlie awoke conscious, and soon began to mend so quickly, that in a few days they were talking of going back to bring Lee-lee in.

CHAPTER XXI.

Lee-lee brought to the Lake—Charlie's Recovery—Final Departure from the Lake.

THE question of getting the infirm old gin across the desert was a somewhat puzzling one.

Charlie, who was fast gaining strength, proposed making Billy and some of the other blacks carry her by turns on a litter of boughs. Brown reminded him that Stuart had found it impossible to get the natives to go to the eastward, so he did not imagine that they would have any better success.

"We must tie her on to a horse, somehow," said Morton at last. And that was all the conclusion they could arrive at.

Charlie was not yet strong enough to stand a long ride, but he felt sufficiently restored to stay behind with only Billy for a companion. So Brown and Morton went back, Charlie having promised to start Billy to meet them with fresh horses on a day appointed.

Lee-lee was anxiously looking out for them, but seemed greatly astonished at seeing two white men. Brown's height, too, appeared to excite his admiration, as it did that of all the blacks they met.

Morton had brought some powder and shot, procured by opening some of their cartridges, as he thought that if he made the gun alive again Lee-lee would come without any difficulty.

"How strange," said Brown, "that these three white men should have lived so long separated from each other and yet within reach."

"I don't know that," replied Morton. "It's rather hard for a man on foot to get about in this country. Remember we have fresh horses, and know where the water is."

Morton inspected the gun.

"I suppose it won't burst," he remarked.

There was a rude ramrod in it, and with a piece of his handkerchief torn off he proceeded to wipe it out. Then he loaded it, Lee-lee watching with great excitement; the old gin, unconscious of their presence, squatting over the half dead fire.

A crow flew, cawing, overhead and settled on a neighbouring tree. Lee-lee pointed eagerly at the bird. Morton raised the gun and fired.

The crow fell down with an angry caw, and the old gin gave a wild scream and tumbled forward on to the fire.

Lee-lee limped after the bird, and the two white men hauled the gin off the fire, which fortunately was nearly out, and dusted the ashes off her.

"You couldn't possibly have hit her?" said Brown.

"Not unless this old blunderbuss shoots round corners. It's the sudden fright."

They put the old creature in the shade, and then the two friends started for a stroll round the lagoon.

When they returned Lee-lee pointed to the old gin as though highly amused at something. She had solved all the difficulties of transport across the desert. She was dead!

"That start I gave her firing off the gun did it," said Morton, sorrowfully; "but she could not have lived much longer."

They indicated to Lee-lee that they would help him bury the old gin; then they saddled up and rode to Hentig's camp, as Brown wanted to see the place, and Morton to recover the pieces of the old powder-flask, which he had neglected to secure on his first visit. With a tomahawk they re-cut the cross on the tree where the remains of Hentig rested.

They got back to Lee-lee's lagoon soon after dark, and devoted an hour or two to packing up all the curious collection of stuff that had so long been hoarded up.

Next morning they made a very early start, as, the half-caste being quite new to riding, they had to go slow. They camped in the desert that night, and about the middle of the next day met Billy coming along the tracks with fresh horses for them. He reported Charlie as being nearly well and everything being safe at the camp. Late in the afternoon, just after they caught sight of the lake, they heard an outcry behind. Looking around they saw Lee-lee limping back, and Billy, who was laughing loudly, pursuing him. It turned out that Lee-lee got a sudden fright at seeing the great sheet of water for the first time, and tumbled off his horse and tried to run back. He seemed reassured after a while, and went on quietly for the rest of the way. Charlie was up and looking nearly as well as ever, and had a fine meal of fish and ducks waiting for them. Lee-lee seemed surprised at the appearance of still a third white man, but took everything else, including his supper, as a matter of course.

Next morning they went over to the black's camp accompanied by Lee-lee. The young fellow who had been wounded was getting rapidly well, Morton or Brown having attended to him and dressed his wound every day. It was soon evident that there was little or no language in common between the two tribes, with the exception of a few words used nearly everywhere in the interior. They had lived and died year after year unconscious of each other's existence.

"We have accounted now for all of Leichhardt's party but one, and he, I think, must have died when the two were separated from the main party," said Morton.

"He could scarcely have got back to where they were attacked by the blacks in the scrub," replied Brown, "and if he had stuck to his companion they would have found the water together. No, he must have perished at the time."

"Now, how about Lee-lee?"

"I think we will stop here for a bit and let Charlie get quite strong and Lee-lee broken into riding a bit, then we will take him back to the station. What do you think?"

"I think it is a good idea; we go round by the way Stuart went and try and pick up his gold reef."

"Yes. We must find out whether one of these old men knows anything about the hole; they ought to."

"Let's go over and make inquiries this afternoon."

This they did, and found out that one of the old men knew of the hole, and had been there once when a young man. He made no objection to going with them, corroborating in this respect Stuart's journal.

They asked after Onkimyong, but, perhaps on account of their faulty pronunciation, did not at first make themselves understood. At last one of the old fellows recognized the name, and pronounced it after his own fashion. The natives immediately pointed to where the bodies lay in the old camp, and they understood that Stuart's faithful companion had met his fate at the hands of the fierce Warlattas, whom he had so often helped to defeat. Both the men had cherished the hope that he might be one of the survivors, as they would then have taken him with them to show them the exact road Stuart travelled in his vain attempt to get away.

From the old men they tried to obtain a description of Stuart's personal appearance, but beyond that he was tall like Brown and had a gray beard, they could not get much information.

They employed their spare time in rigging up a makeshift saddle for Lee-lee to ride on; meantime he took his riding-lessons on one of theirs, and got on famously. He was very proud of being allowed to fire off his gun two or three times a day, and once succeeded in hitting a bird. The time now drew near for their departure. They could do nothing for the natives, but as their enemies were dead, and they lived in a land of plenty, there was no reason why the tribe should not grow up again if they were allowed to remain long enough unmolested.

The natives remained apathetically watching the whites when they departed. Probably they thought that as they came back once, according to their belief, they would come back again.

The stage to the first water was not a long stage on horseback, so the old man kept up with them easily. He knew nothing beyond the lagoon, however, so he was of no further use to them, and they felt confident that they could follow up Stuart's track from his journal. Next morning they gave him a spare tomakawk they had with them and allowed him to depart. Brown, whom he still considered as "Tuartee", having to promise that he would return.

Lee-lee had got on very well with his first day's journey, and they anticipated having no trouble. He was quick and ready in the use of his hands, and, moreover, he and Billy were beginning to understand each other, so they hoped soon to get his history in full. As they had dry country ahead of them, scantily watered, they spelled a couple of days at the white lagoon as they christened it, on account of the milky appearance of the water.

The first day's journey was through the wearisome desert scrub described by Stuart. They calculated what a long day's march on foot would be, but when they had covered that distance there was no sign of the salt lake.

"These salt lakes have no tributaries running into them," said Brown; "they are just depressions, with the surrounding country sloping into the basin. We might be within a quarter of a mile of it and miss it."

"We must find this one at anyrate, if we have to go back and camp for a week at that lagoon," replied Morton.

"Well, it's still two or three hours off sundown, and we have plenty of water for to-night. Suppose you go north and I go south. Charlie and the boys stop here and keep a fire going with plenty of smoke, so that we can get the straight bearing to the camp if either of us drops on it."

"Agreed. North is your lucky cardinal point, so I will take the south."

They started in different directions, while Charlie and Billy took the packs off the horses, and tied them up to trees with their saddles still on, for there was no feed.

Morton went on south for nearly an hour without meeting with any change. He went east and west for short distances as he returned, but was unsuccessful in coming upon any clue to the situation of the salt lake.

Brown was equally unfortunate, until, just as he was on the point of turning back, the unmistakable smell of burning scrub-wood struck on his nostrils.

"It can't be from the camp," he thought; "what little wind there is comes from the north."

He pushed on, and in a few minutes came to an open area, and before him lay the salt lake.

There was a broad belt of mud surrounding a centre of clear water, on which a varied lot of wild fowl, including black swans and wild geese, were swimming. On the slope descending to the edge of the mud there was good short grass growing, and at no distance away he saw the uppiled earth indicating a native well. He rode over to it, and dismounting found a fair supply of water in it. It was slightly brackish, but would do well enough for their horses, being what is generally known as "good stock water".

He next looked all round the lake for the fire which he had smelt, and presently detected the smoke a short way off, stealing out of the edge of the scrub.

"Perhaps it's those six Warlattas," he thought, "and they might be saucy seeing me alone."

He unslung his rifle from his saddle, and advanced with the bridle of his horse on his arm.

CHAPTER XXII.

Another Remnant—An Exodus—Search for the Gold Reef
and its Discovery.

AS he neared the spot he saw two or three
dark figures spring up, as though they then
first noticed him. Fearful that they would run
away, he called to them and held up one hand.
Presently an old man came to the edge of the
scrub. He peered at Brown from under his
hand, for the afternoon sun was in his eyes; then
he burst into a shout of "Tuartee! Tuartee!" so
like the blacks at the lake that Brown thought
some of them must have followed him. This, of
course, he knew to be almost impossible, and as
they were evidently of a friendly disposition, he
walked boldly up. There were only five blacks
in all, the old man and four youths. The young
fellows hung back, but the old man laughed
and stroked Brown affectionately, murmuring
"Tuartee" all the while. There was no doubt
that this was another wretched remnant of the
tribe formerly camped at the mountain, who
had escaped alive from the murderous attacks of
the Warlattas. Stuart would have lived affectionately in the remembrance of those who were
old enough to remember him as their deliverer
on two occasions from their enemies.

It was getting late, however, and Brown told

them he would come back after the sun went
down, and left them, and rode hastily to camp.
It did not take long to replace the packs on the
horses, and by dusk they were all at the lake.
The horses drank the water freely, and were soon
enjoying the young grass. The number of the
blacks had been augmented by two gins, who
had been digging roots on the other side of the
lake when Brown first appeared.

"I've another brilliant idea," said Brown, when
they had finished their meal.

"Let's have it," replied Morton.

"These poor beggars have evidently sought
refuge in this howling wilderness from the Warlattas. As things go, I should not think it
was a very choice place of residence—they look
miserable enough."

"I know what you are going to propose," interrupted Morton. "Get them on to the lake
and let them mate up with the others."

"Exactly. I think it feasible enough; we shall
have to make this our headquarters while we
hunt up that reef. We are not pressed for time
nor rations, thanks to the game at the lake."

"And we sha'n't find that reef in a day, either,"
returned Morton. "We'll sleep on the idea."

Next morning Morton proposed an amendment.
Before the blacks left (if they could induce them
to do so), they should get the old man to guide
them to the soakage spring where Stuart camped
the night after he found the reef. This would

probably be on the usual route travelled by the
blacks, and would considerably contract the area
of their search. While this was going on, Billy,
who had learnt a little of the lake language,
would explain to the natives the advantage of
the change.

" We seem to be constituting ourselves a kind
of special providence for this part of the world,"
said Morton, as he finished.

" We have plenty of time to go to the spring
to-day, if we can make the old fellow understand
what we want."

This they did after some trouble, but it was
evident the native did not enjoy the idea of going
in that direction. However, as the two whites
started with him he finally consented. When
about what they considered half-way, Morton
and Brown parted, Brown going on with the
blackfellow, and Morton intending to devote a
few hours to searching around and then returning to the salt lake. He found no indications,
however, to reward his trouble.

Brown turned up early the next day, the old
fellow having travelled sturdily. He had found
the spring well supplied with fresh water, but
had vainly tried to get anything out of his guide
of a heap of white stones anywhere in the neighbourhood of the track they followed. However,
Brown thought by the formation of the country
about the spring that they could trace the line
back.

"How have you got on with these fellows with regard to an exodus. This old fellow knows all about the lake, but I don't think he has been there."

"Oh, Billy has turned out a splendid orator. He has been gesticulating to them, and fired their imaginations with his descriptions of thousands of wild ducks and millions of fish," said Charlie.

"Now, who is to go back and introduce them to their future companions?"

"I'm all right now," returned Charlie; "Billy and I will shepherd them across."

"It's a good road all the way, I think you will manage it," replied Morton. "How about Lee-lee?"

"We must take him with us when we go out reef-hunting. He might run away if left by himself here," said Brown.

"He is a pretty cute fellow and will help us, if we make him understand what we are looking for. Our camp and horses will be safe enough all day; for, one way and another, the district is getting pretty well depopulated."

The arrangements were so decided on, and the next morning, under convoy of Charlie and Billy, the survivors of the mountain tribe departed for the promised land flowing with birds and fish. After their custom the gins were loaded up with what little camp furniture they possessed, while the lordly male strode along with nothing but a boomerang and a small throw-ing-stick, without which no self-respecting black-fellow would be seen.

Charlie, however, equalized matters by putting what he could on one of the pack-horses, and giving the gins a chance.

Morton, Brown, and Lee-lee set out in the opposite direction. The first day they exhaustively searched for some distance on either side of the track taken by Brown and the old man, but reached half-way to the spring without find-out anything, and returned to the salt lake. Next day Brown proposed that they should go straight to the spring and work back. This they did, taking a pack-horse with rations, and leaving a note for Charlie in a conspicuous place, lest they should be detained and he should come back before they did.

The spring was at the foot of a small hillock strewn with granite boulders. They turned out the horses and started on foot to try and follow the line of country whereon rock was visible on the surface. They managed with great care to keep to it until it was time to return. Next morning they took their horses and rode out to where they had left off. In the middle of the day they turned out for a spell, having been encouraged by finding occasional belts of quartz and slate crossing the granite formation.

As they were smoking after their meal, Lee-lee, who was sauntering about, came back, and pointing on ahead, indicated that a heap of white

stones was there. Both men got up, and in a
few steps saw an outblow of quartz about a
hundred yards away. Hastening to it, they were
soon busy breaking stones and investigating.

They soon found that they had struck Stuart's
reef, or an outcrop on the same line. The stone
appeared to the finders fabulously rich, some of
it being powdered throughout with gold.

"Well, I suppose there's a fortune or two there,"
said Brown when their inspection was over.
"But it's in a deuce of an outlandish place."

"Wonder how far we are across the border
into Western Australia?"

"A good way, I expect; but we will keep the
reckoning very carefully as we go back."

"We have got all we want now; we will pick
out the best of the specimens and take them
with us."

"Yes; and go straight back to the salt lake
and wait for Charlie."

Picking out the richest and smallest specimens,
they packed them on the pack-horse and struck
in for the salt lake on a compass line. This gave
them the bearing from the salt pan, and was all
they wanted to find the place again.

Charlie did not return for a couple more days,
but as they had instructed him to take things
easy, they did not feel anxious.

He had taken his convoy safely to the lake,
and duly introduced the survivors of the two
tribes. Billy and he waited a day to make sure

that amicable relations were properly established and had then returned, everything being peaceful and satisfactory.

Another start was now made for the spring, Brown, Billy, and Lee-lee going straight there with the pack-horses, and Morton taking Charlie round by the reef to show him the rich find.

From the top of the hillock at the back of the spring the country looked scrubby, waste, and desolate; but the outlook was not extensive, and they could see nothing of the mountain they were making for. It behoved them, then, to be very careful, for the country ahead was evidently very dry, and the direction to the creek with the brackish water in it, of the vaguest.

They had a good many things at stake, the safety of Stuart's journal containing the solution of the Leichhardt mystery, and the knowledge of the gold reef. They did not, then, wish to meet with any disaster on their homeward way.

"This is not an exciting sort of road," said Brown, as they turned from fruitlessly scanning the ocean of dull gray tree-tops, "but I think it is a little superior to that abominable desert."

"Yes, we'll patronize this track if ever we come back here; and I suppose we shall come some day to sink on that reef, and see if it goes down."

"If that is the only big show, the gold will be pretty dear before we get it home; but if there is plenty more about, you will soon see a road out here and a township too."

"Go on. A railway, and those gas-lamps and bridges you reported seeing in the scrub."

"Why not? Both you and I have seen those things spring up like magic in Australia, before now."

"Well, I hope our luck will stick to us to-morrow and see us on to that creek."

CHAPTER XXIII.

The Dry Creek—Brown has a Solitary Camp—A Mysterious Light.

IT is unfortunate," said Morton the next morning when they were preparing to start, "that Stuart did not give a description of the creek, or of the water where they camped the next night."

"Yes, it's rather a game of blindman's-buff, for they may have gone north or south of the direct line."

"How far do you make it to the mountain direct?" asked Brown.

A rough chart, compiled every night by dead reckoning, had been kept since they started, and Morton had been working it up the night before.

"Over one hundred miles; and if it is scrub all the way, with sand underfoot, equal to one hundred and fifty."

"No good, then, our striking straight for the mountain and trusting to chance for finding water on the way?"

"Too risky altogether. We must find this brackish creek somehow."

"We can't get back the way we came to the lake, for the last water we camped at must be bone dry by this time."

"How about going round by Hentig's grave if we are beaten back utterly?"

"Yes, as a last resource, we might try that. Lee-lee could not help us, for I don't think he ever stirred far from that lagoon where you found him."

"Let us trust to our lucky star and get on anyhow," returned Morton as he swung himself into his saddle, and they were soon filing slowly through the scrub.

The scrub consisted of mulga and a dense undergrowth of lancewood, so that the progress made was very slow. Moreover, it was a difficult thing to keep a straight course amongst so many obstacles. With the exception that the scrub was sometimes denser than usual they experienced no change, until about two o'clock, when they emerged into a small open space, and Charlie exclaimed that they had come to a grave-yard.

The clearing they had entered was a white clay flat, sparsely grown over with spinifex, and covered with ant-hills about three feet in height, bearing a startling resemblance to the headstones of graves.

The party halted, partly to discuss their movements, and partly to have something to eat. Morton, who finished first, mounted his horse and rode in a southerly direction, telling the others he would be back directly.

"The scrub is thinner to the south-east," he said when he came back, "and beyond I can see another flat like this. I vote we shift our course for a few miles. This change of country may mean that the creek is somewhere about here."

Brown agreed to this, and Morton went ahead.

Passing through a belt of scrub they came to another flat like the one they had left, but somewhat larger. From this they passed through thinner belts of scrub until the flat became continuous, still, however, covered with the ant-hills.

Presently Morton pointed ahead, and a line of creek gums of no great height was now visible. The creek was bordered by a sandy flat with some coarse grass on it. The water-course was shallow, crossed here and there by bars of sand-stone rock; but it was as dry as though water had never been in it for years. It ran easterly.

"This is a lively look-out," said Morton. "Shall

we follow this creek down, or camp here and one go up and one go down the creek?"

"If we find nothing wet we must retreat to the spring to-morrow morning."

"We must; but if we all go on down the creek and find nothing wet, as you express it, we shall be too far to retreat."

"Does not this creek come from much the same direction we came from?" asked Charlie.

"If it keeps the same course as it runs here, it does," replied Morton.

"We are just as likely to find water up as down," went on Charlie; "but if we follow it up we shall be getting nearer to the spring instead of away from it, and if we don't find water can easily cut across to it."

"And have much better travelling-ground along this flat," said Brown. "Charlie, my boy, we shall make a first-class bushmen of you before we get home. A1, copper-fastened at Lloyds."

Charlie's suggestion was unanimously accepted, and the party turned up the creek, making for the westward again. Morton elected to follow the bed of the creek, whilst the others kept as straight a course as they could along the flat.

The creek still continued dry, and no birds of any kind were visible—a bad sign. Like most creeks running through the level interior it gave indications every now and then of running out altogether. At last, however, it grew narrower and deeper, and Morton saw a group of gum-trees ahead, somewhat taller than those lining the banks. There was a small bar of rocks across the channel, and when he rode over this he saw a pool of water before him fringed with green reeds. The water looked strangely clear as he rode down to it; his horse put his head down to drink, but lifted it at once with a dissatisfied snort.

"I guess what's up," thought Morton, dismounting. He stooped and lifted some water to his lips with his scooped hand.

"Bah!" It was salter than brine.

Remounting, he rode up the bank and called to the others, who were visible slightly ahead. They waited when they saw him riding towards them.

"I think we had better ride straight for the spring," he said; "there's water down there, but it's salter than the Pacific Ocean."

"We have good travelling along here," replied Brown. "I think we ought to keep on here as far as we can and then strike off for the spring. It doesn't much matter about water now, for the spring can't be many miles off."

"You follow the creek, then, for a bit; you seem luckier than I am. It does not much matter about the water, as you say, but I should like to know whether there was any fresh water in it as well as salt."

Brown went off to the creek and they once

more started, until Morton calculated that a short three miles through the scrub which was running parallel to them would bring them to the spring. He shouted to Brown and fired his revolver, and when Brown joined them they turned off and reached the spring at dusk.

"Back for the first time," said Morton, as they unpacked at their old camp. "I wonder how many times we shall have to return here."

"Lucky we have such a good camp to stand by us," answered Brown. "We can always get from here to the lake."

The next thing to consider was their movements for the morrow. Morton suggested that perhaps the clay formation altered the conditions of the creek, and below, the water, if not fresh, was at least only brackish.

"I doubt it," said Brown; "these clay formations generally carry salt."

"One of us had better take Billy and a couple of horses packed with water. Let Billy go about twenty miles, then, whoever goes on, give his horse a couple of bags of water and hang the others up on the branch of a tree against his return."

"That's the only safe way," replied Brown. "Who is to go, you or I?"

"You're the lucky one."

"No. You found Lee-lee; let's toss up."

"That's all right, but where's the coin?"

"Rather good," laughed Brown. "Men with

a rich reef in their possession and can't raise a copper to toss with."

"We must shake in the hat," replied Morton.

He tore up a leaf of his note-book, made a mark on one scrap, doubled them up and shook them together in his hat.

Brown drew the marked paper, and chose to go.

"Don't run away with my share of the reef while I am away," he said as he got on his horse early the next morning, and, followed by Billy driving the two horses, was soon lost to sight in the scrub.

"We may as well go out and amuse ourselves at the reef," said Morton; "we can do nothing until he comes back."

They saddled up, and spent the best part of the day in knocking stones out and breaking them, returning in the evening with a few extra rich specimens to add to those they already had.

"If we show these specimens when we get home, won't somebody suspect, and follow our tracks back?" said Charlie.

"If we are fools enough to show them," replied his cousin; "but we'll take all sorts of good care that we don't, until we are ready to come out ourselves, and have pretty well located the place. If Brown does not turn up before morning, we will go out again to-morrow and see if we can trace the reef any further."

Billy turned up with the two horses just at dusk. He had accompanied Brown some miles beyond where they turned back, but there had been no change in the creek as far as he went.

Brown, meanwhile, kept on down the creek after parting with the blackboy. It continued enlarging, and contracting again, in the eccentric manner of an inland water-course, but there was no sign of water, fresh or salt.

The silence, lifelessness, and the gloomy neighbourhood of the scrub on one side of him naturally affected his spirits, and when night fell the sense of loneliness was increased. As it was useless going on in the dark, he determined to give his horse a few hours' rest and then go back.

"The moon rises at twelve o'clock," he thought; "if I start then, I shall get back to the waterbags by daylight."

He short-hobbled his horse, sat down at the foot of a tree with his saddle at his back, and lit his pipe. The great stillness of the desert surrounded and oppressed him with the intensity of its silence. Not a leaf rustled, not a night bird could be heard; the jingle of his horse's hobblechain, and his munching as he cropped the grass, was a welcome sound in that dreary waste. No one knows what a companion a horse is until he has passed a few solitary nights in the uninhabited bush of the interior. Gradually Brown felt sleep stealing over him.

"I can afford to doze," he thought. "I'm pretty uncomfortable, so I sha'n't sleep long."

His head fell back on his saddle, and he was soon fast asleep. He awoke suddenly, feeling stiff and unrefreshed. Springing to his feet, he listened for the sound of his horse; but everything was still.

"What a fool I was to go to sleep!" he thought. "I expect my old prad has made back up the creek, and I shall have to stump it to the camp. Wonder what the time is."

He took his watch out of his pouch, and, the starlight not being strong enough, struck a match. Instantly he was agreeably startled by a loud snort of surprise close to him, and his horse, who had been lying down asleep, got on his legs and shook himself. Brown felt so relieved that he went over and patted and stroked him.

"I thought you had left me in the lurch, old fellow," he said, as he slipped the bridle over his head, for it was nearly midnight, and he thought he might as well make a start. As he stood up after stooping to take the hobbles off, his attention was attracted by a brightness in the eastern sky. "Moon rising," he thought, and led his horse to the tree where his saddle was.

He saddled his horse and was about to mount, when he noticed that the sky was no brighter, and the glow was reddish in colour.

"Moon's rather long-winded," he muttered, and

stood there watching for its appearance; but it obstinately refused to appear.

CHAPTER XXIV.

Fire to the East—Brown returns to the Spring—More
Dry Creeks Discovered.

BROWN stood patiently waiting for some minutes, and then the truth struck him. It was not the moon rising; it was a bush fire, a long distance away.

"Deuced queer," he thought, as he took out his compass, struck a match, and took the bearing of the glow. "It's too far for me to do anything, even if I felt so inclined, which I don't. Hullo! what's this?"

A bright light suddenly gleamed through the trees a little to the south of the other. This, however, was the moon in reality, and Brown turned his willing horse's head towards home, marvelling much at what he had seen.

"Fires travel any distance in this unoccupied country," he thought, "and that one may have come a hundred miles or more."

He reached the water-bags at sunrise, and gave his horse their contents, then, having strapped them on to his saddle, rode on and arrived at the camp at the spring about noon.

Morton could only account for the fire in the same way that Brown did; that it must have travelled a long distance, and that its presence did not denote the existence of water. On his part Morton was able to inform him that they had found another outcrop of the reef that morning, nearly a quarter of a mile to the south, and it appeared as rich as the one they had discovered first.

The waste ahead, however, still sternly confronted them.

"I wonder whether there is another creek further south that this one runs into," said Morton; "or there may be one it joins to the north."

"Very likely; this creek that has been humbugging us does not look to me like a main one, It nearly lost itself several times yesterday, and when I camped it looked very sick."

"We can easily settle the question in a day; to-morrow one go north and one south, as before."

"May I go this time?" said Charlie.

"You go north, and I'll go and crack stones at the new reef," returned Brown.

So it was settled, and they spent a lazy afternoon.

In the morning the two started in opposite directions, and Brown went off to inspect the new find.

Charlie, having been strictly cautioned to trust to his compass only, went due north, and for ten or twelve miles was surrounded by scrub. Then he emerged in a strip of open country, and to his great joy saw creek timber ahead. This watercourse was quite different to the one they had been on—it was more like a chain of shallow lagoons, but all were dry and parched. Charlie followed it for some distance, but there was no sign of moisture, and, elated at having something to report, he made his way back to the spring. Strange to say, when Morton came in he too had found a similar creek to the south, but also waterless. Brown worked out the courses on a bit of paper.

"It strikes me," he said, "that these two creeks, if they run on as they were running where you struck them, must junction in with the creek I was on, not many miles below where I camped."

"Supposing we split up," said Morton. "Say you and Charlie, with half the spare horses, follow down the creek he found, and I and the boys will follow down the one I found, with the rest of the horses. We shall meet at the junction, if your theory is correct. The party who gets there first to wait for the others."

"But supposing there is no water in either of the creeks?"

"We can get back here."

"If your creek junctions in above ours, or *vice versa*, how is the party who arrives at the lower junction to know that the other party is waiting at the upper one?"

"Hum!" said Morton; "that rather capsizes the notion. But I think we can fix it by running the creek up and down a bit."

"Well, I'm willing," returned Brown. "I don't think we are such duffers as to miss each other if we get anywhere within a few miles."

In the morning the plan mooted was carried out, and they left the spring, as they hoped, for good that journey. The creek Brown and Charlie followed proved to be very serpentine in its course. When they stopped for a mid-day spell Brown worked out the dead reckoning, and came to the conclusion that although they had come over fifteen miles in distance, they had not made more than ten in a direct course. Still the creek, on an average, was bearing in towards the other one, and they reckoned they must strike it late in the afternoon.

As they went on the flat grew wider and the empty water-holes further apart, but everything bore the look of a prolonged drought. At four o'clock they sighted the other creek ahead, but there were no signs of the others.

"Wonder how your cousin got on?" Brown said to Charlie. "Hurrah! there he is!" he returned, as a horseman came into sight riding down the bank of the old creek.

Morton pulled up when he caught sight of them, and waited.

"Any water?" he asked when they came up.

"Not a drop. I don't think there has been

any in it since the time of Noah's flood. How did you get on?"

"There was no water in the creek we followed, but there is a decent hole where it junctions with this one, about two miles up from here."

"Salt?"

"No, quite drinkable—a slight sweet taste about it."

"I expect there's more water in it than when Stuart was here: these holes get salter as they dry up. Do you think it is the hole he was at?"

"I think it must be," returned Morton, as they turned and rode up the creek. "We ought to be able to get through to the mountain now, even if we don't come across that clay-pan."

"That's good news, at any rate. Did you see anything of that fire?"

"There appears to be a heavy bank of smoke to the eastward, but we must try and find a tree this evening to have a look-out from."

The camp was a fairly good one, although the grass was somewhat dry. After some searching Brown and Morton found a gum-tree which they could climb, but it was not of a sufficient height to afford them a good view of the surrounding country. They made out, however, that an extensive bush fire was raging to the eastward, and when it fell dark the glow was plainly visible. Brown said it was not as bright as when he saw it, as though the fire was now working away from them.

The following day they started on a straight course for the mountain on the creek, and rode the whole day through a barren region of scrub. That night the horses had to be tied up to trees, for there was neither grass nor water for them. However, they felt sure of arriving at the creek the next day.

"We ought to be getting to that big plain pretty soon," said Morton in the morning, as they were making an early start. "That is, if our reckoning is anyway near the mark."

They had scarcely been travelling an hour, when they suddenly rode from the scrub on to the plain, and before them in the distance, with a black haze of smoke as a background, was the mountain they were making for. The fire was seemingly beyond the mountain, as the plain, although covered with dry grass which would have burnt freely enough, had not been burnt.

Once out of the scrub they travelled more rapidly, and in the afternoon once more camped at the base of the mountain. All the eastern side of the creek was burnt bare, and when they ascended the hill they could see that the fire had ravaged most of the spinifex scrub and burnt up the country to the north. The outlook was even drearier than before, for the heat and flames had scorched the leaves of the low trees, and nothing but an expanse of dead foliage was beneath them.

Fortunately there was good feed for their

horses on the bank of the creek and the islands
in its bed, and as the last two days had been
rather severe on them, they decided to rest for
a few days and inspect the surrounding country,
although it held out little inducement. However, they preferred stopping at where they
were to going back to their old camp at the
lagoons, where probably all the grass was burnt.
The first thing to do was to jot down the whole
of their course since leaving the lagoons and
correct it, which they were now able to do, as
they had arrived back at a known point. They
found that the dead reckoning had been very
well kept, and that their work closed in a satisfactory manner.

An excursion down the creek on the following
day convinced them that it ran out and was
hopelessly lost in the sandy scrub that stretched
south and east. Next morning Morton was up
early at break of day, and climbing up the hill to
reach the summit before sunrise, which is the
best time to see long distances. To the east the
fire was still burning in the distance, but was
evidently now in a dying state. Morton had his
glasses with him, and commenced to carefully
scan the country. At last his attention became
fixed on one particular spot to the south. He
took a compass-bearing and descended the hill.
The others were up, and about to commence
breakfast.

"I've spotted that rock hill," said Morton.

"What! The one Stuart says the old black-fellow told him about?"

"I think so. You can't pick it out with the naked eye, but with the glasses I can make it out quite distinctly. A brown naked cone rising out of the scrub."

"How far away is it?"

"Not more than fifteen miles, I should say. I wonder that none of the niggers were able to take Stuart to it."

"Do you intend going?"

"We may as well. I should like to know all about the place before we go home."

"Well, I'm with you, old man."

Next morning they started on Morton's compass-bearing. The distance was about what he judged, and they made a very fair course.

The rock, surrounded by a small area of open country, rose in a round-topped peak to an altitude of about one hundred and fifty feet. The granite sides were smooth and naked, and the two white men, after hanging their horses to a small cork-tree, climbed to the summit. Brown, who had been in Western Australia before, had seen these granite formations peculiar to that colony, but to Morton they were a new phenomenon. From the top they had a good clear view all round. Scrub, east and south, still stretched before them. Presently they both at the same time noticed a clear space west of south, in which there was a sparkle like a reflection from the sun. Morton turned the glasses on it.

"Salt lake," he said, after a pause.

Brown took the glasses and looked.

"Yes, another salt lake, there's no doubt. We'll take the bearings and apparent distance; it's just as well to have all these things down."

"Not worth while going over to it," said Morton.

They descended the hill and rode round it to see if there were any of the holes on the base of the mound, such as are often found. In this case there were two or three, but all small and dry.

"I don't see any good in going into that scrub to the east," said Morton as they rode home; "we'll make a start the day after to-morrow."

Brown agreed with him, and they reached camp in good time.

CHAPTER XXV.

The Attack by the Surviving Warlattas—Death of Lee-lee —The Last of the Cannibals.

NEXT morning was a lazy one. About eleven o'clock Morton, who was talking to Brown under the shade of a tree, proposed that they should kill a few hours by a ride up the creek, and called to Billy to bring up a couple of

horses. Charlie, who was in the tent, sung out
in reply that he had gone hunting with Lee-lee,
as their clothes were lying on the ground,—for a
blackfellow always likes to strip whenever he
gets a chance.

"I will go and get the horses, they are just
down the creek," said Charlie, when he heard
what his cousin wanted.

He picked up two bridles and went off down
the creek.

Brown and Morton put their saddles down in
readiness.

The horses were not far, and Charlie soon came
back leading two. He had almost reached the
camp when a shrill yell of terror made them all
start.

Out from the forest came Billy, racing and
shouting, and behind him limped Lee-lee. There
was no need to ask what it meant; behind them,
in close pursuit, came other dark forms with up-
raised spears.

"Those Warlattas!" yelled Brown, as he and
Morton sprang for their rifles. Charlie was
transfixed with surprise. Two of the cannibals,
with their spears up, were now close to the
fugitives, the others pressing on so eagerly that
they did not see the white men. It all seemed
to Charlie to pass like a flash. The spears flew,
and the rifles cracked so closely one after the
other that it sounded almost like one report.
Down went Billy and Lee-lee, and the two War-

lattas behind them pitched forward headlong on
the ground. Startled by the firearms the others
halted, turned and fled. But the breech-loaders
spoke once more, and one Warlatta fell with a
broken leg, and the other dropped in a heap and
lay quiet, with a conical bullet between his shoulders.

"Quick! not one must get away," said Morton,
and he and Brown snatched the bridles from
Charlie's hand, and jumping on bare-backed,
galloped like avenging furies after the two re-
treating survivors. "Look after Billy!" yelled
Morton to Charlie, as he urged his excited horse
along.

The blacks fled into the forest, but the cover
came too late for them, with two of the best
riders in central Australia thirsting for their
blood. Charlie, as he went down to Billy, saw
his cousin race up to one man and they disappeared between the trees, but the report of a
revolver immediately after told its tale. Next
minute came two more pistol-shots from the
direction Brown had gone.

Billy had sat up by the time Charlie reached
him; he had been speared in the leg, but poor
Lee-lee was dead. The spear of the Warlatta
had pierced his heart.

Morton's and Brown's voices were now heard
coming back. They pulled up at the wounded
savage, and Morton slipped from his horse.
Charlie turned his head away, for he guessed

what was going to happen. No quarter for the cannibals. He heard the revolver ring out, and knew that Lee-lee was avenged.

His cousin came up, leading his horse and putting his revolver back in his pouch. Both men were flushed, and their eyes still blazed with the fierce light of conflict.

"Poor Lee-lee!" said Brown, as they stood beside his body. "We seem to have been his evil genius."

"We've been the evil genius of the Warlattas, thank goodness," said Morton grimly. "They're all wiped out now, however."

The tragedy affected them all strongly. The unfortunate half-caste meeting his death in such an unexpected manner, when all seemed safe and at peace, was sad.

Billy, however, demanded their attention. Fortunately the spear was not a barbed one, and had only gone into the fleshy part of his thigh. It was soon extracted, the wound bound up, and he was made as comfortable as possible.

Billy explained that he and Lee-lee were on their way home when they saw the Warlattas, who had evidently been stalking them for some time. Had Billy been armed with firearms he might have frightened them away; but as he had nothing but a tomahawk, he thought discretion the better part of valour and ran for it, forgetting in his excitement that Lee-lee was lame and could not keep up with him.

They buried the last poor relic of Leichhardt's doomed party at the foot of the mountain, but the bodies of the Warlattas were left to the crows and hawks.

"Perhaps it is all for the best, sad as it seems," said Morton. "Those six devils could not keep their lust for murder under, and but for this row we might not have run across them. Then they would have gone to the lake again and finished their villainous work."

"I wonder where they got their weapons from."

"There must have been some left in the bottle tree camp in the basalt. We did not look about much, if you remember."

"Well, that's the end of it all, I suppose."

"Unless somebody comes across Lee-lee's brothers or sisters amongst the tribe to the north."

The party perforce had now to remain where they were until Billy was able to ride again, and a dull time it was. A trip to the hot swamp showed them that, during their absence at the lake, the water had subsided and the swamp become so dry that the fire had ravaged it, burning the ragged, inflammable bark of the trees, and licking up the reeds surrounding the lakelet, which was now but a surface of cracked mud.

"There is one question that always worries me," said Brown, as they came to the spot where the Warlatta track led into the basalt rocks. "Do you think that Murphy was compelled to join in their cannibal feasts?"

"I have thought of it too," replied Morton, "and have come to the conclusion that he was not. At least, while he retained his reason. When we saw him, you know, he was nearly blind, and his mental faculties almost gone. My reason for this is the anchor we found cut on the tree at the lagoons; I daresay there were more, and there were numberless marks of the others. There was an ample game supply up and down that creek, and I believe he spent most of his time there hunting, until he became too infirm to leave the cave."

"I am glad you think that, as I am of the same belief. I think any white man, no matter how slow his intellect, would prefer death."

"Still, cases have been known where men have been maddened by starvation in an open boat at sea; but in this case he would not have been desperate with hunger. No, I think, and am glad to think, that he had no part in their evil doings or rites until he was irresponsible for his actions."

"They would not have allowed him to go with them on their raids for fear of his escaping. Evidently they regarded him as a sort of fetish."

They dismounted and hung their horses to a tree, and went a short distance amongst the rocks. As they advanced all signs of a track

(M 64)   O

disappeared, for the place became one jumbled mass of huge boulders piled on top of one another, rough as a rasp underfoot, and baking hot from the vertical sun. What with the natural heat of the day and the radiation from the rocks, they were soon glad to turn back to where they had left their horses.

" No wonder poor Stuart, barefooted and alone, could not make his way any distance," remarked Morton.

" I wonder what would have happened had he met the Warlattas?"

" He had established a good funk amongst them, and so he might have routed them. But if they had killed him, I swear a good many would have lost the number of their mess first."

" It always makes me feel sad when I think of such a man being forced by fate to spend his life amongst savages."

Billy's wound, like the flesh of most black-fellows, was rapidly healing, but he was not yet able to ride. The shadow cast on their spirits by the murder of poor Lee-lee, rendered them all anxious to be on the move and leave the ill-omened camp behind them. The weather had been continuously fine ever since they left. That night, however, a black thunder-storm gathered up, and towards evening the heavens were overcast and the sky was one constant blaze of lightning, and a continuous mutter of thunder sounded from all points. Every preparation had been made, and they watched with interest the mustering of the storm spirits.

"I believe it's going to be one of those dry dust-storms after all," said Brown.

To the east every blaze of light now showed a low black cloud approaching.

"It's the wind coming," said Morton, "bringing all the ashes from the burnt country; we shall be smothered with dust and charcoal."

Even as he spoke there came a blinding glare of white light, accompanied by a crash of thunder that seemed to shake the hill to its foundation. A rush of cold wind, bearing dust and ashes on its wings, swept the camp and nearly carried away the tent. Then the rain fell in one heavy downpour. For nearly an hour the deluge kept up, the continuous flashes making it as bright as day, the constant roar and rattle of the thunder never ceasing. Then the tumult died away in the west, the stars peeped out, and the tropical storm was over. Next morning the sky was clear and the air fresh and pleasant.

"I'm hanged if I can stop in camp any longer," said Morton. "Billy, if you don't get that 'mundoee' of yours well soon, we'll go away and leave you here."

Billy looked rather askance at the threat, until he realized that Morton was joking.

Brown, who had been surgeon, said: "I think we can rig up a sling or cradle for his leg soon, so that he will be able to travel short stages."

"I'm glad to hear it. That thunder-storm must have put water into the rock-holes at the granite rock. What do you say to a ride there and then on to the salt lake we saw at a distance?"

"Right; it will kill time. But we'll start to-morrow; let the ground dry up a bit. We'll experimentalize on a cradle for Billy's leg to-day."

CHAPTER XXVL

Visit to the Southern Salt Lake—The Future of the Interior—A False Alarm—Departure.

THE cradle promised to be a success; so the next morning, taking some rations in case they had to camp out, Brown and Morton left for the rock. The ground was still somewhat soft, but not enough to impede their travelling, and they reached the granite rock early. As they had the bearing of the salt lake they did not climb the rock again, but rode round the base to see if the holes were full. They were all brim-full, the sloping rock above acting like the roof of a house in catching and shedding the rainfall. They then struck out for the salt lake, which they reached about one o'clock, passing through sandy country all the way. The lake was much larger than the one they had camped at to the north, but the surrounding country was barren and grassless. Few signs of the former presence of the natives were visible, and no indication of a well having been dug. Evidently the soil was so impregnated with salt that not even brackish water could be obtained.

"What a real desert!" said Brown, gazing round on the dreary scene.

"Yes, it's about as hopeless a looking picture as one could find anywhere, at present. And yet, if the artesian water is found to extend throughout the interior, it will change the whole face of the Australian earth in time. This spinifex would not grow here, but that the climate is so arid that nothing else will grow, and this beastly stuff can thrive without any rain at all. No, burn this scrub off, or clear it somehow, and, with a good supply of artesian water, there are a hundred and one payable products one could grow here."

"You're an optimist, and an enthusiast at that."

"I am as regards the future of Australia. I believe the end of the coming century will see it settled from east to west throughout."

"If one could fill up all the dry creeks and lagoons we have passed with your artesian water, we might modify the severity of the climate."

"Yes. Now, let's have a ride round this inland sea in miniature."

"It smells like the sea, at anyrate; I bet that

water in there is concentrated brine. How about all this saline country?"

"It has been proved successfully that the date-palm will thrive on the shores of these salt lakes, so they need not be quite barren."

Nothing of any interest was to be seen, and they retraced their steps to the granite rock, where they watered their horses. As there were still a couple of hours of daylight, they started back for their camp.

"Fancy if we had left the camp like this, forgetting all about those six Warlattas hanging about. What a massacre they would have had!" said Brown, as they rode on.

"Yes, it makes me shudder to think of our carelessness; for we ought to have remembered there was danger to be expected from them."

When it fell dark they found themselves still some three miles from home, and the darkness somewhat retarded them in the scrub. Suddenly, when nearing the mountain, a rifle-shot was heard ahead, followed soon after by a different report, like that of a shot-gun.

"Good God! what can be up?" exclaimed Morton.

Both men fired their revolvers as a signal that they were near, and pushed on as hastily as they could. As soon as the open country was reached they galloped straight for the camp. Everything appeared peaceful enough, and Charlie seemed surprised at their hasty approach.

"What were you firing at?" asked Morton, rather crossly, for no man likes to be flurried by a false alarm.

"Well, I don't know exactly," replied Charlie. "I had given you up for to-night, and was sitting out here with Billy, when he called out that there was something moving on the rocks over there. I looked, and could indistinctly make out some dark figure moving about, so I challenged; getting no answer, I fired my rifle in the air. Whatever it was they started away, but in a few minutes came back again, so I fired the shot-gun at them and they departed. Billy called out they were 'Jinkarras!' and covered his head with the blanket, and I expect he has it there now."

"What were they like?" asked Brown.

"It was too dark to see, but they were certainly not natives, unless we have run across a race of dwarfs."

Billy, on being induced to take his head from underneath the blanket, asserted stoutly that they were Jinkarras they had seen; that he ought to know, as when he was a child he had been carried off by one in the night.

"How did you get back, Billy?" asked Morton.

Billy commenced a long rambling yarn about waking up to find himself being carried along by a short, hairy man with red eyes; but his tale ended somewhat lamely, for his next remembrance was of finding himself in the familiar family camp, with his mother administering severe slaps with the small end of a nulla-nulla. Still he persisted in his statement that there were Jinkarras, and that they lived underground.

"I shouldn't wonder," suddenly exclaimed Brown, "if this legend of the Jinkarras, which is common all over the central portion of Australia, was not a surviving tradition, much distorted, of our dear old friends the devil worshippers."

"Not at all unlikely. We will run this particular brand of Jinkarra to earth in the morning," answered Morton.

Charlie was out before breakfast to inspect the ground where he had seen the figures in the night; but beyond a few good-sized boulders, which he was certain he had not fired at, he failed to discover any marks of a nocturnal visit.

Morton went out after breakfast, and immediately saw what had caused the alarm. He called Charlie over and pointed the tracks out to him.

"This is a regular pad for the rock-wallabies," he said. "Only it has been covered up by the burnt ashes of the grass. They were coming in last night to feed on the young grass on the bank of the creek, just springing after the rain. I suppose some of them hopped on to these boulders."

This explanation failed to satisfy Billy, who was still convinced that the Jinkarras were about, and was now anxious to get away.

They devoted themselves to finishing the sling for his leg, and made him take a short ride two or three times, to get accustomed to it and find out if it hurt him.

It was with feelings of great thankfulness that they at last got ready to make a final start and leave the place which had grown so wearisome to them. For the sake of making it easy for Billy, they intended to take two days on the journey to the lagoons, so they camped the first night on the creek above what had been the hot swamp.

The next night they reached the familiar camp at the lagoons, and now felt that they were finally on the homeward track. They had made a rude pair of crutches for the black boy, and he was now able to limp about on, what he called, his "waddy-mundoees".

As a matter of satisfaction they spelled a day, for although the grass had all been burnt by the fire, there was still good feed on the banks of the lagoons. This day was devoted to thoroughly examining the trees up and down the creek, and they were able to partly confirm their conjectures about Murphy, by finding the anchor marked on several more trees.

The thunder-storm had filled the small hole they stopped at when they first sighted the plain and the great limestone rock, so they made a short stage there to give Billy every chance. From what they remembered of the nature of the country, there was not likely to be any water retained along the scrub track.

They were all on the look-out during the next morning for the spot where they first encountered the Warlattas. When they reached it they found that the corpse was gone, the six men despatched having seemingly done their duty and taken it on to the burying-place.

"I suppose," said Brown, "that it was only men of importance amongst them that they took the trouble to carry all this way. What did they do with the others?"

"I forgot all about it," exclaimed Charlie.

They both looked at him in surprise.

"When I was down that hole, the first one, not the tunnel affair, I saw some bones and skulls amongst the boulders. I think it was that which frightened Billy so. I could only see a few, but there might have been thousands, for everything was smothered with mud and our candles did not give much light."

"At that rate, the rank and file were thrown into the boiling spring when they pegged out," said Morton.

"Seemingly so," answered his friend. "But we must push on, we have a good step ahead of us."

The horses went merrily along the cleared track, and as Billy showed no signs of fatigue they made capital progress. As they anticipated, the cleared track led them straight on to the open patch of downs country where the cemetery was. A great surprise awaited them. The fire had swept up from the south, and the whole country was black. More than that, the fierce flames had attacked the dry boughs forming the scaffolds whereon the dead bodies had been bestowed, and now, all that was to be seen were half-charred bones lying here and there.

"It seems that Fate meant to destroy all traces of the Warlattas in one act," said Morton, as they sat on their horses and gazed at all that was left of the cemetery of the cannibals.

"How was it this never happened before?" remarked Brown.

"I don't understand. They must have kept it burnt down short every year, and neglected it for some reason. However, I'm not sorry, for if this country extends any distance south I shall take it up."

"Well, let's get to camp before it's dark. There will be enough grass unburnt about the water-hole for our horses to-night."

This proved to be the case, and the cheery camp-fire was soon blazing brightly and everybody chatting in good spirits.

"If you think seriously of taking up this bit of country, we might as well explore it to-morrow now we are here. The horses will be better for the rest, for remember, as far as we know, there is not a drop of water between here and the station—a good hundred miles," said Brown.

"That thunder-storm has been along here by the look of it. It should have put some water in some of those clay-pans we passed."

"Thunder-storms are mighty uncertain things to trust to. They generally fall, as a rule, just where they are no good to any one. We must travel, when we start, as though it was dry the whole way, although I think with you that we shall find water."

"As it now stands," said Morton, drawing his blanket over his shoulders, "the only real evidence we have to show that the Warlattas ever existed, is this cleared road in the scrub."

"And the wound in Billy's leg," murmured Charlie, drowsily.

CHAPTER XXVII.

Home Again.

THE trip next morning was a promising one. The creek kept a continued and well-watered course for about fifteen miles, running through well-grassed downs country all the way. The place was burnt black with the fire, but that did not hide the value of the country. Gradually the scrub, which they had lost sight of for some time, closed in on both sides, and it was evident that the creek would soon run out, once it entered the scrub. They were back in camp in time to take a short ride up the creek, and ascertain that there was nothing worth troubling about in that direction. Brown fossicked out the remains of the brandy when they had finished their meal.

"Now, then," he said, when they had all put some in their pannikins, "we must christen the new run. What's it to be? You speak first, Charlie."

"Warlatta Downs."

"Good!" said Morton; "we can't better that. Here's good luck to Warlatta Downs."

"Now for the gold reef," said Brown.

There was silence whilst each thought of a suitable name.

"Suppose we call it after Stuart, who was really the first finder of it."

"The Stuart Reef, then, and here's to his memory."

They drank the toast in silence.

"That reminds me," remarked Brown, "that portion of the diary relating the finding of the gold reef must be carefully eliminated from the original journal and our copies."

"We'll set about it now, to make sure. We can restore it at any time when needful; meantime we don't want anybody to jump our claim."

They soon had the work finished, and the part taken out was carefully put away.

"One more night and home," said Charlie delightedly the next morning as they mounted.

"I never thought so much of the old station before."

The belt of scrub had still to be passed which had proved such a terror on their outward way. Sorely did they miss the well-cleared track of the Warlattas. Luckily the thunder-storm had extended most of the way, and they reached home by easy stages.

"We have not lost a single horse in spite of all the dry and desert country we have negoti-ated," said Morton, as they rode over the familiar ground some miles away from the station.

"No; that's something to boast of. Those long spells we had at different places were the salvation of our nags," replied Brown.

Their return that night caused great excite-ment on the station. The men had been getting impatient and anxious, and were thinking of starting on their tracks to see if they had come to grief.

Every Australian bushman knows the story of Leichhardt, and when the men heard that the mystery of his fate and of those who accom-panied him had been at last solved, they felt that a reflected glory was shed on all connected with the station.

Billy had a great reception from his country-men camped about the station. He exhibited his wound, and let it be generally understood that he had wiped out the Warlatta tribe single-handed, although they were all giants over seven feet high. Fortunately he knew nothing of the gold reef, so was not able to dilate on that; but the story of the lake and the caves there lost nothing by telling, but he quite forgot to mention his fright in the underground tunnel.

The news of their successful trip and interesting discoveries was soon flashed along the overland telegraph-line. It was enthusiastically received by some and scornfully doubted by others, as is usual in these cases. Brown regretted that they had not had a camera, and brought a few pictures back with them; but as the authenticity of the documents have been since universally admitted, the scoffers are confounded.

As yet they are awaiting their time before returning to open up the reef, which they anticipate will be found to be joined by a line of auriferous country with the rich gold discoveries lately made in Western Australia.

THE END.

Blackie & Son's
Illustrated Story Books

HISTORICAL TALES BY
G. A. HENTY

With Buller in Natal: or, A Born Leader. With 10 page Illustrations by W. RAINEY, R.I., and a Map. 6s.

The heroic story of the relief of Ladysmith forms the theme of one of the most powerful romances that have come from Mr. Henty's pen. When the war breaks out, the hero, Chris King, and his friends band themselves together under the title of the Maritzburg Scouts. From first to last the boy scouts are constantly engaged in perilous and exciting enterprises, from which they always emerge triumphant, thanks to their own skill and courage, and the dash and ingenuity of their leader.

"A glowing tale of heroism."—*Eastern Daily Press.*
"Just the sort of book to inspire an enterprising boy."—*Army and Navy Gazette.*

— In the Irish Brigade: A Tale of War in Flanders and Spain. With 12 page Illustrations by CHARLES M. SHELDON. 6s.

The hero is a young officer in the Irish Brigade, which for many years after the siege of Limerick formed the backbone of the French army. He goes through many stirring adventures, successfully carries out dangerous missions in Spain, saves a large portion of the French army at Oudenarde, and even has the audacity to kidnap the Prime Minister of England.

"A good subject made good use of."—*Spectator.*
"Should be popular among boys of all ages."—*Yorkshire Post.*

— Out with Garibaldi: A Story of the Liberation of Italy. With 8 page Illustrations by W. RAINEY, R.I., and two Maps. 5s.

Mr. Henty makes the liberation of Italy by Garibaldi the groundwork of an exciting tale of adventure. The hero is an English lad who joins the expedition and takes a prominent part in the extraordinary series of operations that ended in the fall of the Neapolitan kingdom.

"A first-rate story of stirring deeds."—*Daily Chronicle.*
"Full of hard fighting, gallant rescues, and narrow escapes."—*Graphic.*

From *WITH BULLER IN NATAL*

By G. A. Henty. 6s. (see page 1)

ᴍ 683

"ONE OF THE BOERS HELD UP HIS RIFLE WITH A WHITE FLAG
TIED TO IT"

G. A. HENTY

No Surrender! A Tale of the Rising in La Vendée. With 8 page Illustrations by STANLEY L. WOOD. Crown 8vo, cloth elegant, olivine edges, 5s.

This story tells of the heroic defence of La Vendée against the overwhelming forces of the French Republic. The hero, a young Englishman, joins the Vendéans and renders them invaluable services as leader of a band of scouts.

"Vivid tale of manly struggle against oppression."—*The World.*

"Crammed . . . with fighting, sieges, assaults, and escapes."—*Educational Times.*

– Both Sides the Border: A Tale of Hotspur and Glendower. With 12 page Illustrations by RALPH PEACOCK. 6s.

The hero casts in his lot with the Percys and becomes esquire to Sir Henry, the gallant Hotspur. He is sent on several dangerous and important missions in which he acquits himself with great valour.

"With boys the story should rank among Mr. Henty's best."—*Standard.*

"A vivid picture of that strange past . . . when England and Scotland . . . were torn by faction and civil war."—*Onward.*

– Through Russian Snows: or, Napoleon's Retreat from Moscow. With 8 page illustrations by W. H. OVEREND. 5s.

Julian Wyatt becomes, quite innocently, mixed up with smugglers, who carry him to France, and hand him over as a prisoner to the French. He subsequently regains his freedom by joining Napoleon's army in the campaign against Russia.

"The story of the campaign is very graphically told."—*St. James's Gazette.*

"One of Mr. Henty's best books, which will be hailed with joy by his many eager readers."—*Journal of Education.*

"Is full of life and action."—*Journal of Education.*

– The Young Colonists: A Tale of the Zulu and Boer Wars. With 6 Illustrations by SIMON H. VEDDER. 3s. 6d.

The story of two English lads who serve the British force as guides against Cetewayo, are present at the disaster of Isandula, help to defeat the Zulus at Ulundi, and afterwards fight through the campaign against the Boers.

"No boy can read this book without learning a great deal of South African history at its most critical period."—*Standard.*

G. A. HENTY

Under Wellington's Command: A Tale of the Peninsular War. With 12 page Illustrations by WAL PAGET. 6s.

In this stirring romance Mr. Henty gives us the further adventures of Terence O'Connor, the hero of *With Moore at Corunna*. We are told how, in alliance with a small force of Spanish guerillas, the gallant regiment of Portuguese levies commanded by Terence keeps the whole of the French army in check at a critical period of the war, rendering invaluable service to the Iron Duke and his handful of British troops.

"Will be found extremely entertaining."—*Daily Telegraph.*

"An admirable exposition of Mr. Henty's masterly method of combining instruction with amusement."—*World.*

"Humour, adventure, and hard fighting."—*Navy and Army.*

– At Aboukir and Acre: A Story of Napoleon's Invasion of Egypt. With 8 page Illustrations by WILLIAM RAINEY, R.I. 5s.

Shortly before the battle of the Nile, Edgar Blagrove, the son of an English merchant in Alexandria, saves the life of a young Bedouin chief. The two boys become inseparable, and on the arrival of the French, Edgar makes common cause with the Bedouins against the invader. He afterwards enters the British navy as a midshipman, and as interpreter to Sir Sydney Smith assists in the defence of Acre.

"The boys who are so fortunate as to get the book as a Christmas present will enjoy many hours of supreme delight, and will learn almost unconsciously much that is worth knowing."—*Manchester Guardian.*

"A thoroughly patriotic story, with brisk action, and incidents crowding upon each other."—*Tatler.*

– With Cochrane the Dauntless: A Tale of his Exploits. With 12 page Illustrations by W. H. MARGETSON. 6s.

It would be hard to find, even in sensational fiction, a more daring leader than Lord Cochrane, or a career which supplies so many thrilling exploits. The manner in which, almost single-handed, he scattered the French fleet in the Basque Roads is one of the greatest feats in English naval history.

"As rousing and interesting a book as boys could wish for."—*Saturday Review.*

"This tale we specially recommend."—*St. James's Gazette.*

"We honour the author of *With Cochrane the Dauntless* as the head of his profession."—*National Observer.*

"Full of thrilling adventure, as well as of historical and biographical information." —*Glasgow Herald.*

G. A. HENTY

Won by the Sword: A Tale of the Thirty Years' War. With 12 page Illustrations by CHARLES M. SHELDON. 6s.

In this story Mr. Henty completes the history of the Thirty Years' War, the first part of which he described in *The Lion of the North*. His hero, the son of a Scottish officer who, at an early age, comes under the notice of the great Turenne, and is placed on his personal staff, has ample opportunity for gratifying his love of hazardous enterprises and adventures.

> "As fascinating as ever came from Mr. Henty's pen."—*Westminster Gazette.*

> "Full of sieges, of the smoke, the din and the dust of battle."—*Standard.*

-- By England's Aid: or, The Freeing of the Netherlands (1585-1604). With 10 page Illustrations by ALFRED PEARSE, and 4 Maps. 6s.

Two English lads go to Holland in the service of one of "the fighting Veres". After many adventures one of the lads finds himself on board a Spanish ship at the defeat of the Armada, and escapes from Spain only to fall into the hands of the Corsairs. He is successful, however, in getting back to Spain, and regains his native country after the capture of Cadiz.

> "Boys know and love Mr. Henty's books of adventure, and will welcome his tale of the freeing of the Netherlands."—*Athenæum.*

> "Mr. Henty can give you the sense of battle in the veins."—*National Observer.*

> "Geoffrey's adventures will impart to the tale that element which lays hold of the boy reader."—*Christian Leader.*

- By Right of Conquest: or, With Cortez in Mexico. With 10 page Illustrations by W. S. STACEY, and 2 Maps. 6s.

The conquest of Mexico, by a small band of resolute men under the magnificent leadership of Cortez, is rightly ranked amongst the most romantic exploits in history. With this as the groundwork of his story, Mr. Henty has interwoven the adventures of an English youth, Roger Hawkshaw, the sole survivor of the good ship *Swan*, which had sailed from a Devon port to challenge the supremacy of the Spaniards in the New World.

> "Mr. Henty's skill has never been more convincingly displayed than in this imirable and ingenious story."—*Saturday Review.*

> "Cleverly written and wonderfully interesting."—*Birmingham Gazette.*

> "A volume full of interest and excitement, which cannot fail to charm its readers."—*Journal of Education.*

"'SILENCE! SIGNORS,' HE SAID IN A LOUD VOICE"

G. A. HENTY

A Roving Commission: or, Through the Black Insurrection of Hayti.

With 12 page Illustrations by WILLIAM RAINEY, R.I. 6s.

The hero of this story takes part in some of the principal engagements in the revolt of the slaves of Hayti against their French masters at the end of last century, and is able to rescue many of the unfortunate French colonists from the infuriated blacks. He also does good service against the pirates who infested the West Indian seas at that period, for which he is rapidly promoted from midshipman to commander.

> "A stirring tale, which may be confidently recommended to schoolboy readers."
> —*Guardian.*
> "A singularly lucky and attractive hero, for whom boy readers will have an intense admiration."—*Standard.*

– Beric the Briton: A Story of the Roman Invasion of Britain. With 12 page Illustrations by W. PARKINSON. 6s.

Beric is a boy-chief of a British tribe which takes a prominent part in the insurrection under Boadicea : and after the defeat of that heroic queen he continues the struggle in the fen-country. Ultimately Beric is defeated and carried captive to Rome, where he succeeds in saving a Christian maid by slaying a lion in the arena, and is rewarded by being made the personal protector of Nero. Finally, he escapes and returns to Britain, where he becomes a wise ruler of his own people.

> "He is a hero of the most attractive kind. . . . One of the most spirited and well-imagined stories Mr. Henty has written."—*Saturday Review.*
> "His conflict with a lion in the arena is a thrilling chapter."—*School Board Chronicle.*
> "Full of every form of heroism and pluck."—*Christian World.*

– The Dash for Khartoum: A Tale of the Nile Expedition. With 10 page Illustrations by JOHN SCHÖNBERG and J. NASH. 6s.

In the record of recent British history there is no more captivating page for boys than the story of the Nile campaign, and the attempt to rescue General Gordon. For, in the difficulties which the expedition encountered, and in the perils which it overpassed, are found all the excitement of romance, as well as the fascination which belongs to real events.

> "*The Dash for Khartoum* is your ideal boys' book."—*Tablet.*
> "It is literally true that the narrative never flags a moment."—*Academy.*
> "*The Dash for Khartoum* will be appreciated even by those who don't ordinarily care a dash for anything."—*Punch.*

G. A. HENTY

Bonnie Prince Charlie: A Tale of Fontenoy and Culloden. With 12 page Illustrations by GORDON BROWNE. 6s.

The hero, brought up by a Glasgow bailie, is arrested for aiding a Jacobite agent, escapes, but is wrecked on the French coast, reaches Paris, and serves with the French army at Dettingen. He succeeds in obtaining, through Marshal Saxe, the release from confinement of both his parents. He kills his father's foe in a duel, and, escaping to the coast, shares the adventures of Prince Charlie.

"Mr. Henty can tell a capital story; but here, for freshness of treatment and variety of incident, he has surpassed himself."—*Spectator.*

"The adventures and incidents throughout are of the most exciting kind, and the interest is never for one moment allowed to flag."—*Literary World.*

"Is most intensely thrilling."—*Pall Mall Gazette.*

– In the Heart of the Rockies: A Story of Adventure in Colorado. With 8 page Illustrations by G. C. HINDLEY. 5s.

The hero, Tom Wade, goes out to his uncle in Colorado, who is a hunter and gold-digger. Going in quest of a gold mine, the little band is overwhelmed by a snow-storm in the mountains, and compelled to winter there. They build two canoes and paddle down the terrible gorges of the Rocky Mountains, and after many perils they reach Fort Mojarve in safety.

"No book will please more than *In the Heart of the Rockies.*"—*Spectator.*

"It is a book to read and to recommend to boys and girls."—*The Observer.*

"It is all life and go and vigour from beginning to end."—*The School Board Chronicle.*

– At Agincourt: A Tale of the White Hoods of Paris. With 12 page Illustrations by WAL PAGET. 6s.

Sir Eustace de Villeroy, in journeying from Hampshire to his castle in France, made young Guy Aylmer one of his escort. Soon thereafter the castle was attacked, and the English youth displayed such valour that his liege-lord made him commander of a special mission to Paris. This he accomplished, returning in time to take part in the campaign against the French which ended in the glorious victory for England at Agincourt.

"Is one of Mr. Henty's best."—*Standard.*

"There is not a better book for boys in Mr. Henty's extensive repertory."—*Scotsman.*

"Cannot fail to commend itself to boys of all ages."—*Manchester Courier.*

Blackie & Son's
Story Books for Boys

G. MANVILLE FENN

Devon Boys: A Tale of the North Shore. With 12 page Illustrations by GORDON BROWNE. 6s.

The scene is laid on the picturesque rocky coast of North Devon, where the three lads pass through many perils both afloat and ashore. Fishermen, smugglers, naval officers, and a stern old country surgeon play their parts in the story, which is one of honest adventure with the mastering of difficulties in a wholesome manly way, mingled with sufficient excitement to satisfy the most exacting reader.

"An admirable story, as remarkable for the individuality of its heroes as for the excellent descriptions of coast scenery and life in North Devon. One of the best books we have seen this season."—*Athenæum.*

– Nat the Naturalist: A Boy's Adventures in the Eastern Seas. With 8 page Pictures by GORDON BROWNE. 5s.

The boy Nat and his uncle go on a voyage to the islands of the Eastern seas to seek specimens in natural history, and their adventures there are full of interest and excitement. The descriptions of Mr. Ebony, their black comrade, and of the scenes of savage life sparkle with genuine humour.

"This book encourages independence of character, develops resource, and teaches a boy to keep his eyes open."—*Saturday Review.*

– Yussuf the Guide: With 6 page Illustrations by J. SCHÖNBERG. 3s.

A lad who has been almost given over by the doctors, but who rapidly recovers health and strength in a journey through Asia Minor with his guardians and Yussuf as their guide. Their adventures culminate in their being snowed up for the winter in the mountains, from which they escape while their captors are waiting for the ransom that does not come.

"This story is told with such real freshness and vigour that the reader feels he is actually one of the party, sharing in the fun and facing the dangers w th them."
—*Pall Mall Gazette.*

From BOY CRUSOES

BY LEON GOLSCHMANN.   3s. 6d.   (See page 12)

"THE BEASTS HURLED THEMSELVES SAVAGELY AGAINST THE DOOR

Dr. GORDON STABLES, R.N.

In Far Bolivia: A Story of a Strange Wild Land. With 6 page Illustrations by J. FINNEMORE, R.I. 3s. 6d.

Life on the beautiful plantation on the banks of the great Amazon flows gently and dreamily on, until the abduction of the heroine by Bolivian savages. Then the stir indeed begins, and the adventures of the rescue-party, in which the heroine's boy cousin and his chum are the moving spirits, are the subject of an enthralling narrative.

"Written in Dr. Gordon Stables' best style."—*Yorkshire Herald.*
"An exciting and altogether admirable story."—*Sheffield Telegraph.*

– Kidnapped by Cannibals: A Story of the Southern Seas. With 6 page Illustrations by J. FINNEMORE. 3s. 6d.

Thinking his father harsh and himself aggrieved, Willie runs away to sea with his friend Harry Blessington. There is a terrible mutiny on board, stirred up by foreign sailors, and in a group of hitherto unknown cannibal islands in the southern seas the boys are cast away. Their strange, wild life and adventures for years among these cannibals are told in most graphic style.

"Exciting and interesting from beginning to end."—*Punch.*
"Full of exciting adventure, and told with spirit."—*Globe.*

Capt. F. S. BRERETON, R.A.M.C.

In the King's Service: A Tale of Cromwell's Invasion of Ireland. With 8 page Illustrations by STANLEY L. WOOD. 5s.

When the Parliamentary army crosses to Ireland young Dick Granville and his cousin Terence join a body of Royalist horse. They take part in the defence of Drogheda, only escaping from the slaughter there by a miracle, and afterwards go through a series of thrilling adventures and narrow escapes in which Dick displays extraordinary skill and resource.

"What boy could wish for a better story?"—*Eastern Morning News.*

– With Rifle and Bayonet: A Story of the Boer War. With 8 page Illustrations by WAL PAGET. 5s.

On the outbreak of war Jack Somerton volunteers as a despatch rider. After rendering signal service, he shares the fate of many another gallant fellow, and is brought as a captive to Pretoria. He escapes in due course, and is fortunate in sharing in the brilliant operations that resulted in the capture of the Free State capital.

"Boys who are not ready to vote this the book of the season do not deserve to be remembered when Boxing-day comes round." *Glasgow Herald.*

LÉON GOLSCHMANN

Boy Crusoes:
A Story of the Siberian Forest. Adapted from the Russian by LÉON GOLSCHMANN. With 6 page Illustrations by J. FINNEMORE, R.I. 3s. 6d.

Every page of *Boy Crusoes* is crammed with adventure of a fresh and peculiarly interesting character. It is a story of two Russian lads who are so deeply impressed by reading *Robinson Crusoe* that they decide to run away from home and have similar experiences. They lose their way in a huge trackless forest, and are kept busy hunting for food, fighting against wolves and other enemies, and labouring to increase their comforts, before they are rescued by chance after some two years' sojourn in the forest.

"This is a story after a boy's own heart."—*Nottingham Guardian.*

ROBERT LEIGHTON

The Thirsty Sword:
A Story of the Norse Invasion of Scotland (1262–63). With 6 page Illustrations by A. PEARSE. *New Edition.* 3s.

The story tells how Roderic MacAlpin, the sea-rover, came to the Isle of Bute; how he slew both his brother, Earl Hamish, and the earl's eldest son, Alpin: how young Kenric became King of Bute, met the sea-rover on Garroch Head, and ended their feud.

"One of the most fascinating stories for boys that it has ever been our pleasure to read. From first to last the interest never flags."—*Schoolmaster.*

MEREDITH FLETCHER

Every Inch a Briton:
A School Story. With 6 page Illustrations by SYDNEY COWELL. 3s. 6d.

This story is written from the point of view of an ordinary boy, who gives an animated account of a young public-schoolboy's life. No moral is drawn; yet the story indicates a kind of training that goes to promote veracity, endurance, and enterprise; and of each of several of the characters it might be truly said, he is worthy to be called, "Every Inch a Briton".

"In *Every Inch a Briton* Mr. Meredith Fletcher has scored a success." —*Manchester Guardian.*

Col. J. PERCY-GROVES

The War of the Axe:
A Story of South African Adventure. Fully Illustrated. 2s. 6d.

Tom Flinders, late of Rugby, sails to rejoin his parents in Cape Colony, goes up country, meets with many experiences, gets mauled by a Cape lion, and finally joins the Cape Mounted Rifles.

"The story is well and brilliantly told, and the illustrations are especially good and effective."—*Literary World.*

FRED. HARRISON

Wynport College: A Story of School Life. With 8 page Illustrations by HAROLD COPPING. 5s.

The hero and his chums differ as widely in character as in personal appearance. We have Patrick O'Fflahertie, the good-natured Irish boy; Jack Brookes, the irrepressible humorist; Davie Jackson, the true-hearted little lad, on whose haps and mishaps the plot to a great extent turns; and the hero himself, whose experiences at Wynport College the story relates.

"Crammed with happy incident."—*Truth.*
"Full of life and adventure."—*Standard.*

W. C. METCALFE

All Hands on Deck! With 6 Illustrations by W. RAINEY, R.I. 3s. 6d.

The story begins with the experiences of eight of the crew and a young lady passenger in an open boat. After many privations they are picked up by the ship *Ariadne*. By a rare combination of circumstances the hero is compelled to assume the command of the *Ariadne*, and navigates the ship safely to Hong Kong, where many happy surprises await him.

"Is such a graphic sea story that the reader almost sniffs the salt breeze of mid-ocean."—*Literary World.*

EDGAR PICKERING

An Old-Time Yarn: Adventures in the West Indies and Mexico with Hawkins and Drake. With 6 page Illustrations by ALFRED PEARSE. 3s. 6d.

The hero sails from Plymouth in the flagship of Master John Hawkins. Divers are the perils through which he passes. Chief of these are the destruction of the English ships by the treacherous Spaniards, the fight round the burning vessels, the journey of the prisoners to the city of Mexico, the horrors of the Inquisition, and the final escape to England.

"An excellent story of adventure. . . . The book is thoroughly to be recommended."—*Guardian.*

WALTER P. WRIGHT

An Ocean Adventurer: or, The Cruise of the Orb. With 4 page Illustrations by PAUL HARDY. 2s. 6d.

There is a fight at the beginning of this story, and treasure at the end of it. From the day he is robbed of his father's priceless secret, through all the desperate adventures of the privateer *Orb*, until he solves the great mystery, Frank Pingle has a thrilling time of it, and so will his reader.

"A breezy and exciting sea story . . . which is full of stirring incidents from beginning to end."—*Court Circular.*

W. O'BYRNE

Kings and Vikings: Stories of Irish History. With 6 page Illustrations by PAUL HARDY. 2s. 6d.

In this inspiring collection the author of *A Land of Heroes* gives us a further series of heroic stories from the ancient annals of Erin. The present volume has all the romantic charm that attaches to the early history of Ireland.

"The stories are full of human interest, and are well told."—*Bradford Observer.*

S. BARING-GOULD

Grettir the Outlaw: A Story of Iceland in the days of the Vikings. With 6 page Illustrations by M. ZENO DIEMER. 3s.

A narrative of adventure of the most romantic kind. No boy will be able to withstand the magic of such scenes as the fight of Grettir with the twelve bearserks, the wrestle with Karr the Old in the chamber of the dead, the combat with the spirit of Glam the thrall, and the defence of the dying Grettir by his younger brother.

"Is the boys' book of its year. That is, of course, as much as to say that it will do for men grown as well as juniors. It is told in simple, straightforward English, as all stories should be, and it has a freshness, a freedom, a sense of sun and wind and the open air, which make it irresistible."—*National Observer.*

C. J. CUTCLIFFE HYNE

The Captured Cruiser: or, Two Years from Land. With 6 page Illustrations by F. BRANGWYN. *New Edition.* 3s. 6d.

The central incidents deal with the capture, during the war between Chili and Peru, of an armed cruiser. The heroes and their companions break from prison in Valparaiso, board this warship in the night, overpower the watch, escape to sea under the fire of the forts, and finally, after marvellous adventures, lose the cruiser among the icebergs near Cape Horn.

"The two lads and the two skippers are admirably drawn. Mr. Hyne has now secured a position in the first rank of writers of fiction for boys."—*Spectator.*

- Stimson's Reef: With 4 page Illustrations by W. S. STACEY. *New Edition.* 2s. 6d.

This is the extended log of a cutter which sailed from the Clyde to the Amazon in search of a gold reef. It relates how they discovered the bucaneer's treasure in the Spanish Main, fought the Indians, turned aside the river Jamary by blasting, and so laid bare the gold of *Stimson's Reef*

"Few stories come within hailing distance of *Stimson's Reef* in startling incidents and hairbreadth 'scapes. It may almost vie with Mr. R. L. Stevenson's *Treasure Island*"—*Guardian.*

From EVERY INCH A BRITON

By MEREDITH FLETCHER.   3s. 6d.   (See page 12)

"DOE, DINAH, DINAH, DINAH, DINAH, DINAH DOE!"

MICHAEL MACMILLAN
Tales of Indian Chivalry. With 6 page Illustrations by PAUL HARDY. 2s. 6d.

Professor Macmillan's great acquaintance with India and Indian literature has placed an almost unequalled store of romance at his disposal; and the present collection of typical stories of Rajput, Moghul, and Mahratta chivalry should win for itself a permanent place among the treasures of the British schoolboy.

"Capital reading for boys."—*Outlook*.

HARRY COLLINGWOOD
The Congo Rovers: A Tale of the Slave Squadron. With 6 page Illustrations by J. SCHÖNBERG. 3s.

The scene of this tale is laid on the west coast of Africa. The hero, after being effectually laughed out of his boyish vanity, develops into a lad possessed of a large share of sound common-sense, the exercise of which enables him to render much valuable service to his superior officers in unmasking a most daring and successful ruse on the part of the slavers.

"No better sea story has lately been written than *The Congo Rovers*. It is as original as any boy could desire."—*Morning Post*.

FLORENCE COOMBE
Boys of the Priory School. With 4 page Illustrations by HAROLD COPPING. 2s. 6d.

The interest centres in the relations of Raymond and Hal Wentworth, and the process by which Raymond, the hero of the school, learns that in the person of his ridiculed cousin there beats a heart more heroic than his own.

"It is an excellent work of its class, cleverly illustrated with 'real boys' by Mr. Harold Copping."—*Literature*.

JANE H. SPETTIGUE
A Trek and a Laager: A Borderland Story. With 4 page Illustrations by PAUL HARDY. 2s. 6d.

A story of pioneer life in South Africa, full of stirring adventure and interesting information about the country. The hero and his sister are cut off from the laager by a party of natives, and are captured by the Kafirs after making a bold stand against overwhelming odds. They escape and reach the laager, which is afterwards fiercely attacked by the natives, but is successfully defended.

"Full of novelty and excitement."—*Spectator*.

Blackie & Son's
Story Books for Girls

KATHARINE TYNAN

Three Fair Maids: or, The Burkes of Derrymore. With 12 full-page Illustrations
by G. DEMAIN HAMMOND, R.I. 6s.

A story of Irish country life. On the initiative of Elizabeth, the eldest and most brilliant of the widowed Lady Burke's daughters, they receive "paying guests" at Ardeelish, the house in which they have lived since Sir Jasper's disinheritance obliged them to give up their great house, Derrymore. Through the "paying guests" Elizabeth and Joan both meet their fates; and the family is reconciled with Uncle Peter Burton, who makes Elizabeth his heiress.

"One of the author's best Irish tales, and everyone knows how charming they can be."—*St. James's Gazette.*

ELIZA POLLARD

The King's Signet: The Story of a Huguenot Family. With 6 full-page
Illustrations by G. DEMAIN HAMMOND, R.I. 3s. 6d.

This story relates the adventures of a noble Huguenot family, driven out of their chateau by the dragoons after the Revocation of the Edict of Nantes. The adventures lead to the battle of the Boyne, and to the happy reunion of the scattered family in Ireland.

"The story is splendidly told, and never drags."—*Pall Mall Gazette.*

CAROLINE AUSTIN

Cousin Geoffrey and I. With 6 full-page Illustrations by W. PARK-
INSON. 3s.

The only daughter of a country gentleman finds herself unprovided for at her father's death, and for some time lives as a dependant upon her kinsman. Life is kept from being unbearable to her by her young cousin Geoffrey, who at length meets with a serious accident for which she is held responsible. She makes a brave attempt to earn her own livelihood, until a startling event brings her cousin Geoffrey and herself together again.

"Miss Austin's story is bright, clever, and well developed."—*Saturday Review.*

ELLINOR DAVENPORT ADAMS

A Queen among Girls. With 6 Illustrations by HAROLD COPPING.

Crown 8vo, cloth elegant, 3s. 6d.

Augusta Pembroke is the head of her school, the favourite of her teachers and fellow-pupils, who are attracted by her fearless and independent nature, and her queenly bearing. She dreams of a distinguished professional career; but the course of her life is changed suddenly by pity for her timid little brother Adrian, the victim of his guardian-uncle's harshness. The story describes the daring means adopted by Augusta for Adrian's relief.

"An interesting and well-written narrative, in which humour and a keen eye for character unite to produce a book happily adapted for modern maidens."—*Globe*.

- A Girl of To-Day. With 6 page Illustrations by G. D. HAMMOND, R.I. 3s. 6d.

"What are Altruists?" humbly asks a small boy. "They are only people who try to help others," replies the *Girl of To-Day*. To help their poorer neighbours, the boys and girls of Woodend band themselves together into the *Society of Altruists*. That they have plenty of fun is seen in the shopping expedition and in the successful Christmas entertainment.

"It is a spirited story. The characters are true to nature and carefully developed. Such a book as this is exactly what is needed to give a school-girl an interest in the development of character."—*Educational Times*.

FRANCES ARMSTRONG

A Fair Claimant. The Story of a Girl's Life. With 6 page Illustrations by G. DEMAIN HAMMOND, R.I. *New Edition.* 3s.

The heroine, when a child, is found deserted in an attic. She is adopted by a wealthy lady, and resides abroad until the death of her benefactress. Thereafter, Olive Bethune comes to England as a governess, and then begins to learn her own strange history. It is a tale of surprising vicissitude, but in the end all the wrongs are pleasantly righted.

"There is a fascination about this story. The splendid character of the heroine, together with the happy manner in which the interest is sustained to the end, combine to make this one of the most acceptable gift-books of the season."—*Church Review*.

G. NORWAY

A True Cornish Maid. With 6 page Illustrations by J. FINNEMORE. 3s. 6d.

The heroine of the tale is sister to a young fellow who gets into trouble in landing a contraband cargo on the Cornish coast. In his extremity the girl stands by her brother bravely, and by means of her daring scheme he manages to escape.

"The success of the year has fallen, we think, to Mrs. Norway, whose *True Cornish Maid* is really an admirable piece of work."—*Review of Reviews*.

From THREE FAIR MAIDS

By Katharine Tynan. 6*s*.   (See page 17)

м 630

"Lord Kinvarra going before her to open the
little Wicket"

ROSA MULHOLLAND (LADY GILBERT)

Cynthia's Bonnet Shop. With 8 full-page Illustrations by G. DEMAIN HAMMOND, R.I. 5s.

Cynthia and her star-struck sister Befind go to London, the former to open the bonnet shop, and the other to pursue the study of astronomy. How both girls find new interests in life, more important even than bonnet shop or star-gazing, is described with mingled humour and pathos in this charming story.

"Just of the kind to please and fascinate a host of girl readers."
—Liverpool Mercury.

- Giannetta: A Girl's Story of Herself. With 6 full-page Illustrations by LOCKHART BOGLE. 3s.

The story of a changeling who is suddenly transferred to the position of a rich English heiress. She develops into a good and accomplished woman, and has gained too much love and devotion to be a sufferer by the surrender of her estates.

ANNIE E. ARMSTRONG

Violet Vereker's Vanity. With 6 full-page Illustrations by G. DEMAIN HAMMOND, R.I. 3s. 6d.

The heroine was an excellent girl in most respects. But she had one small weakness, which expressed itself in a snobbish dislike of her neighbours the Sugdens, whose social position she deemed beneath her own. In the end, however, the girl acknowledged her folly, with results which are sure to delight the reader.

"A book for girls that we can heartily recommend, for it is bright, sensible, and with a right tone of thought and feeling."*—Sheffield Independent.*

ALICE CORKRAN

Margery Merton's Girlhood. With 6 full-page Illustrations by GORDON BROWNE. 3s. 6d.

The experiences of an orphan girl who in infancy is left by her father—an officer in India—to the care of an elderly aunt residing near Paris. The accounts of the various persons who have an after influence on the story are singularly vivid.

"*Margery Merton's Girlhood* is a piece of true literature, as dainty as it is delicate, and as sweet as it is simple."*—Woman's World.*

MRS. R. H. READ

Dora: or, A Girl without a Home. With 6 page Illustrations by PAUL HARDY. 3s. 6d.

The story of an orphan girl, who is placed as pupil-teacher at the school in which she was educated, but is suddenly removed by hard and selfish relatives, who employ her as a menial as well as a governess. Through a series of exciting adventures she makes discoveries respecting a large property which is restored to its rightful owners, and at the same time she secures her own escape.

"*Dora* is one of the most pleasing stories for young people that we have met with of late years. There is in it a freshness, simplicity, and naturalness very engaging."
—*Harper's Magazine.*

MRS. E. J. LYSAGHT

Brother and Sister: With 6 page Illustrations by GORDON BROWNE. 3s. 6d.

A story showing, by the narrative of the vicissitudes and struggles of a family which has "come down in the world", and of the brave endeavours of its two younger members, how the pressure of adversity is mitigated by domestic affection, mutual confidence, and hopeful honest effort.

"A pretty story, and well told. The plot is cleverly constructed, and the moral is excellent."—*Athenæum.*

BESSIE MARCHANT

The Girl Captives: A Story of the Indian Frontier. With 4 page Illustrations by WILLIAM RAINEY, R.I. Crown 8vo, cloth elegant, 2s. 6d.

The ladies and children of an Indian frontier town are carried off by border tribesmen, but through a tribesman to whom the heroine, the daughter of an English officer, had formerly done an act of kindness, they make their escape after many stirring adventures.

"Altogether a capital little book."—*Saturday Review.*

SARAH TYTLER

A Loyal Little Maid. With 4 page Illustrations by PAUL HARDY. 2s. 6d.

This pretty story is founded on a romantic episode of Mar's rebellion. A little girl has information which concerns the safety of her father in hiding, and this she firmly refuses to divulge to a king's officer. She is lodged in the tolbooth, where she finds a boy champion, whom in future years she rescues in Paris, from the *lettre de cachet* which would bury him in the Bastille.

"Has evidently been a pleasure to write, and makes very enjoyable reading."
—*Literature.*

"LEANT HER FLUFFY HAIR AND SOFT CHEEK AGAINST CAROL'S KNEE"

GERALDINE MOCKLER
The Four Miss Whittingtons: A Story for Girls. With
8 full-page Illustrations by CHARLES M. SHELDON. 5*s.*

This story tells how four sisters, left alone in the world, went to London to seek their fortunes. They had between them £400, and this they resolved to spend on training themselves for the different careers for which they were severally most fitted. On their limited means this was hard work, but their courageous experiment was on the whole very successful.

"A story of endeavour, industry, and independence of spirit."—*World.*

ANNE BEALE
The Heiress of Courtleroy. With 8 full-page Illustrations by
T. C. H. CASTLE. 5*s.*

Mimica, the heroine, comes to England as an orphan, and is coldly received by her uncle. The girl has a brave nature, however, and succeeds in saving the estate from ruin and in reclaiming her uncle from the misanthropical disregard of his duties as a landlord.

ALICE STRONACH
A Newnham Friendship. With 6 full-page Illustrations by HAR-
OLD COPPING. 3*s.* 6*d.*

A sympathetic description of life at Newnham College. After the tripos excitements, some of the students leave their dream-world of study and talk of "cocoas" and debates and athletics to begin their work in the real world. Men students play their part in the story, and in the closing chapters it is suggested that marriage has its place in a girl graduate's life.

"Foremost among all the gift-books suitable for school-girls this season stands Miss Alice Stronach's *A Newnham Friendship.*"—*Daily Graphic.*

BESSIE MARCHANT
Held at Ransom: A Story of Colonial Life. With 4 full-page Illustrations by SYDNEY
COWELL. 2*s.* 6*d.*

A story of life in Cape Colony shortly after the discovery of diamonds at Kimberley. How the heroine ransoms her father, but herself falls into the hands of his captors, and how both are eventually rescued, is told in a most graphic and entertaining manner.

"A stirring, graphically-written story."—*Bookman.*

SARAH TYTLER

Queen Charlotte's Maidens. With 3 Illustrations by PAUL HARDY. 2s.

The scene is laid in the White House, a beneficent institution in which a number of orphan girls drawn from the professional classes earn their living by fancy needlework, under the immediate patronage and protection of the Queen. The element of romance creeps in, and one after another of the inmates of the White House leaves its portals for a wider sphere.

"It is a pretty tale in a picturesque setting."—*Western Morning News.*

SARAH DOUDNEY

Under False Colours: A Story from Two Girls' Lives. With 6 page Illustrations by G. G. KILBURNE. 4s.

A story which will attract readers of all ages and of either sex. The incidents of the plot, arising from the thoughtless indulgence of a deceptive freak, are exceedingly natural, and the keen interest of the narrative is sustained from beginning to end. *Under False Colours* is a book which will rivet the attention, amuse the fancy, and touch the heart.

"This is a charming story, abounding in delicate touches of sentiment and pathos. Its plot is skilfully contrived. It will be read with a warm interest by every girl who takes it up."—*Scotsman.*

E. EVERETT-GREEN

Miriam's Ambition. With Illustrations. 2s. 6d.

Miriam's ambition is to make someone happy, and her endeavour carries with it a train of incident, solving a mystery which had thrown a shadow over several lives. A charming foil to her grave elder sister is to be found in Miss Babs, a small coquette of five, whose humorous child-talk is so attractive.

"Miss Everett-Green's children are real British boys and girls, not small men and women. Babs is a charming little one."—*Liverpool Mercury.*

EMMA LESLIE

Gytha's Message: A Tale of Saxon England. With Illustrations. 2s. 6d.

We get a glimpse of the stirring events taking place at that period; and both boys and girls will delight to read of the home life of Hilda and Gytha, and of the brave deeds of the impulsive Gurth and the faithful Leofric.

"This is a charmingly told story. It is the sort of book that all girls and some boys like, and can only get good from."—*Journal of Education.*

Blackie & Son's Finely Illustrated Books for Children

SHEILA E. BRAINE

The Princess of Hearts.

With Frontispiece in colour and 70 Illustrations by ALICE B. WOODWARD. F'cap 4to, cloth elegant, gilt edges, 6s.

Take a Princess, a mysterious Duchess, frog and fairy combined, an Ogre—Mugwump by name, malevolent by nature,—a palace inhabited by the Royal Family of Hearts, and a marsh gay with "Winking Marybuds", and alive with Queer Folk. Add a Contradicter and a peppery Scullerymaid, and we have the essential ingredients of this delightful fairy tale.

"A valuable addition to fairy-tale lore, worthily illustrated by Alice B. Woodward." —*Queen*.

JUDGE PARRY—WALTER CRANE

The Story of Don Quixote.

Retold by His Honour Judge PARRY. Illustrated by WALTER CRANE, with 11 coloured full-page Plates, 19 half-page Plates, a Title-page, and Cover. Royal 8vo, cloth, 6s.

This is an abridgment of Cervantes' immortal book, and an endeavour to write in simple narrative form the adventures of Knight and Squire with as much of the wisdom and humour of their discourse as will be within the grasp of the younger generation of readers.

"The book world of children is the richer for a new and attractive version of an immortal story."—*Manchester Courier*.

MRS. PERCY DEARMER

Roundabout Rhymes.

With 20 full-page Illustrations in colour by Mrs. PERCY DEARMER. Imperial 8vo, cloth extra, 2s. 6d.

A charming volume of verses and colour pictures for little folk—rhymes and pictures about most of the everyday events of nursery life.

"The best verses written for children since Stevenson's *Child's Garden*. Altogether we commend this book as a very charming piece of design, and more especially as verse, touched with a great deal of insight and humour, yet perfectly simple and amusing."—*The Guardian*.

HARRY B. NEILSON

Droll Doings.
Illustrated by HARRY B. NEILSON, with Verses by the Cockiolly Bird. Twenty-three full pages and 18 vignettes in full colour. Royal 4to, picture boards, cloth back, 6s.

In *Droll Doings* Mr. Harry B. Neilson has produced a picture-book of striking originality and irresistible humour. He shows us our most familiar friends in the animal world engaged in the same pursuits, playing the same tricks, and involved in the same scrapes, as ordinary human girls and boys. The Cockiolly Bird pipes a merry note, and his song will add not a little to the enjoyment of the happy youngsters into whose hands this delightful volume may come.

"Distinctly clever and funny."—*Record.*
"A really delightful picture book."—*Gentlewoman.*

MABEL E. WOTTON

The Little Browns.
With Frontispiece in colour and more than 80 Illustrations by H. M. BROCK. F'cap 4to, cloth elegant, gilt edges, 6s.

The little Browns are a delightful set of youngsters whom the upbringing of an inert mother and faddist father has rendered more than usually individual and self-reliant. During their parents' absence they extend hospitality to a stranger under the belief that he is their unknown and long-expected uncle from Australia. Two of the little girls make the discovery that the supposed uncle is really a burglar who is in league with the new man-servant, and by their courage and childish resource outwit him.

"Young readers will find *The Little Browns* irresistibly attractive."—*Observer.*

EDITH KING HALL

Adventures in Toyland.
With 4 page Pictures printed in colour, and 70 black-and-white Illustrations throughout the text, by ALICE B. WOODWARD. Square 8vo, cloth extra, 2s. 6d.

In daylight, and when the shop was open, the Toys neither stirred nor spoke; but when the shutters were shut and all the grown-ups had gone, the Toy World became alive. This happy discovery was made by little Molly, who found that her friend the Marionette could tell wonderful tales about Toyland, and the quaint adventures of the Toys.

"One of the funniest as well as one of the daintiest books of the season. The adventures are graphically described in a very humorous way."—*Pall Mall Gazette.*

From DROLL DOINGS

By Harry B. Neilson. 6s. (See page 26)

"WE'LL SEE IF THOSE SILLY OLD BEES ARE ALIVE"

(Reduced from a full-page colour picture)

OUR DARLING'S FIRST BOOK

Bright Pictures and Easy Lessons for Little Folk.
Quarto, 10⅛ inches by 7¾ inches, picture boards, 1s.; cloth, gilt edges, 2s.

An interesting and instructive picture lesson-book for very little folk. Beginning with an illustrated alphabet of large letters, the little reader goes forward by easy stages to word-making, reading, counting, writing, and finally to the most popular nursery rhymes and tales.

"The very perfection of a child's alphabet and spelling-book."—*St. James's Budget.*

JENNIE CHAPPELL

Mignonne:
or, Miss Patricia's Pet. With a Frontispiece and 20 Illustrations by PAUL HARDY. 2s.

Miss Patricia has long desired in vain to adopt the child, who is left in her charge. A severe illness deprives the little girl of all recollection of the past, and causes a report of her death to be sent to her father, who is abroad. Mignonne's eventual restoration, with recovered memory, to her family, brings the story to a happy end.

"There can be no more desirable book for schoolroom or fireside."
—*Whitehall Review.*

A. B. ROMNEY

Little Village Folk.
With 37 Illustrations by ROBERT HOPE. 2s. 6d.

A series of delightful stories of Irish village children. Miss Romney opens up a new field in these beautiful little tales, which have the twofold charm of humour and poetic feeling.

"A story-book that will be welcomed wherever it makes its way."—*Literary World.*

ALICE TALWIN MORRIS

The Elephant's Apology.
With over 30 Illustrations by ALICE B. WOODWARD. Square 8vo, decorated cloth, 2s. 6d.

Why it was very necessary and how his Worship the Mayor was graciously pleased to accept the apology is set forth in pleasant words and pictures. Other animal stories follow, but it would take more than a page to tell of half the delights to be found in these pretty tales both by children and by their elders.

"We have not seen a more charming or dainty book for children."
—*Pall Mall Gazette.*

STORIES BY GEORGE MAC DONALD
(NEW AND UNIFORM EDITION)

A Rough Shaking. With 12 page Illustrations by W. PARKINSON. Crown 8vo, cloth elegant, 3s. 6d.

Clare, the hero of the story, is a boy whose mother is killed at his side by the fall of a church during an earthquake. The kindly clergyman and his wife, who adopt him, die while he is still very young, and he is thrown upon the world a second time. The narrative of his wanderings is full of interest and novelty, the boy's unswerving honesty and his passion for children and animals leading him into all sorts of adventures. He works on a farm, supports a baby in an old deserted house, finds employment in a menagerie, becomes a bank clerk, is kidnapped, and ultimately discovers his father on board the ship to which he has been conveyed.

At the Back of the North Wind. With 75 Illustrations by ARTHUR HUGHES, and a Frontispiece by LAURENCE HOUSMAN. Crown 8vo, cloth elegant, 3s. 6d.

"In *At the Back of the North Wind* we stand with one foot in fairyland and one on common earth. The story is thoroughly original, full of fancy and pathos."—*The Times.*

Ranald Bannerman's Boyhood. With 36 Illustrations by ARTHUR HUGHES. Crown 8vo, cloth elegant, 3s. 6d.

"Dr. Mac Donald has a real understanding of boy nature, and he has in consequence written a capital story, judged from their stand-point, with a true ring all through which ensures its success."—*The Spectator.*

The Princess and the Goblin. With 30 Illustrations by ARTHUR HUGHES, and a Frontispiece by LAURENCE HOUSMAN. Crown 8vo, cloth elegant, 3s. 6d.

In the sphere of fantasy George Mac Donald has very few equals, and his rare touch of many aspects of life invariably gives to his stories a deeper meaning of the highest value. His *Princess and Goblin* exemplifies both gifts. A fine thread of allegory runs through the narrative of the adventures of the young miner, who, amongst other marvellous experiences, finds his way into the caverns of the gnomes, and achieves a final victory over them.

The Princess and Curdie. With Frontispiece and 30 Illustrations by HELEN STRATTON. Crown 8vo, cloth elegant, 3s. 6d.

A sequel to *The Princess and the Goblin*, tracing the history of the young miner and the princess after the return of the latter to her father's court, where more terrible foes have to be encountered than the grotesque earth-dwellers.

"GEE UP, DADDY!"

(*Reduced from a full-page colour picture*)

NEW "GRADUATED" SERIES

With coloured frontispiece and black-and-white illustrations

NO child of six or seven should have any difficulty in reading and understanding *unaided* the pretty stories in the 6*d.* series. In the 9*d.* series the language used is slightly more advanced, but is well within the capacity of children of seven and upwards, while the 1*s.* series is designed for little folk of somewhat greater attainments. If the stories are read *to* and not *by* children, it will be found that the 6*d.* 9*d.* and 1*s.* series are equally suitable for little folk of all ages.

"GRADUATED" STORIES AT A SHILLING

Betty the Bold. By ELLINOR DAVENPORT ADAMS.
Jack of Both Sides. By FLORENCE COOMBE.
Do Your Duty! By G. A. HENTY. *New Edition.*
Tony's Pains and Gains. By W. L. ROOPER.
Terry. By ROSA MULHOLLAND (Lady Gilbert).
The Choir School. By FREDERICK HARRISON.
The Skipper. By E. CUTHELL.
What Mother Said. By L. E. TIDDEMAN.
Little Miss Vanity. By Mrs. HENRY CLARKE.
Two Girls and a Dog. By JENNIE CHAPPELL.
Miss Mary's Little Maid. By ELLINOR DAVENPORT ADAMS.

"GRADUATED" STORIES AT NINEPENCE

That Boy Jim. By Mrs. HENRY CLARKE.
The Adventures of Carlo. By KATHARINE TYNAN.
The Shoeblack's Cat. By W. L. ROOPER.
Three Troublesome Monkeys. By A. B. ROMNEY.
The Little Red Purse. By JENNIE CHAPPELL.
Put to the Proof. By Mrs. HENRY CLARKE.
Teddy's Ship. By A. B. ROMNEY.
Irma's Zither. By EDITH KING HALL.
The Island of Refuge. By MABEL MACKNESS.

"GRADUATED" STORIES AT SIXPENCE

Bravest of All. By MABEL MACKNESS.
Winnie's White Frock. By JENNIE CHAPPELL.
Lost Toby. By M. S. HAYCRAFT.
A Boy Cousin. By GERALDINE MOCKLER.
Travels of Fuzz and Buzz. By GERALDINE MOCKLER.
Teddy's Adventures. By Mrs. HENRY CLARKE.
Sahib's Birthday. By L. E. TIDDEMAN.
The Secret in the Loft. By MABEL MACKNESS.
Two Little Friends. By JENNIE CHAPPELL.
Tony's Pets. By A. B. ROMNEY.
Andy's Trust. By EDITH KING HALL.

NEW CHILDREN'S PICTURE BOOKS

IN DOORS AND OUT | STORY-BOOK TIME

Pictures and Stories for Little Folk. Each contains 38 colour pages,
over 40 full-page black-and-white Illustrations, and a large
number of Vignettes. Quarto, 10⅛ inches by 7¾ inches, picture-
boards, 2s. 6d. each; cloth, gilt edges, 3s. 6d. each.

MOST attractive books of stories, rhymes, and pictures for little readers.
There is no double page without a picture, and the many colour pages
in bright tints will prove specially acceptable to young folk. Santa Claus
could bring no more welcome gift than one of these pretty volumes.

ONE SHILLING SERIES

Quarto, 10⅛ inches by 7¾ inches

My Very Best Book.
Arm-chair Stories.

My Very Own Picture Book.
Cosy Corner Stories.

BRIGHT and amusing picture books for the little folk. Each volume
contains over twenty full-page drawings by eminent artists, and a large
number of smaller illustrations. The cover, and no fewer than twenty
pages, are printed in colour.

SIXPENNY SERIES

Quarto, 10⅛ inches by 7¾ inches

Smiles and Dimples.
Little Bright-Eyes.
For Kittie and Me.
As Nice as Nice Can Be.
Round the Mulberry Bush.
Little Rosebud.

For My Little Darling.
For Dolly and Me.
My Own Story Book.
Play-time Pictures.
Bed-time Stories.
For Little Chicks.

IT may confidently be said that these are the most attractive picture books
ever published at the price. Each book contains an average of six full-
page illustrations, a large number of vignettes, and seven pages in colour.
The cover designs, also in colour, are extremely attractive, the text is printed
in bold type, and the stories and rhymes that form the letterpress are bright
and humorous.